Climate-Adaptive Design in High Mountain Villages

Drawing from the unique context and climate of the Himalaya, this book highlights several innovative design interventions, shaped by a myriad of social, cultural, environmental, and political factors that have been employed in villages to combat climate change.

Climate-Adaptive Design in High Mountain Villages focuses on Ladakh, an outpost on the front lines of climate change, and the region's creative responses to the pressing issues of food security, water management, energy efficiency, design aid, and material resources in the Anthropocene. These strategies – from artificial glaciers to tree armor – showcase the breadth of creative solutions already underway. In doing so, the research addresses the broader concept of climate-adaptive design and how it informs the disciplines of architecture, landscape architecture, and urban planning.

An ideal read for academics, researchers, and students in these fields, this book presents a focused investigation into climate-adaptive strategies that could provide transferable solutions for the rest of the world.

Carey Clouse is an Associate Professor of Architecture and Landscape Architecture at UMass Amherst. She is the author of *Farming Cuba: Urban Agriculture from the Ground Up* (2014), the recipient of a Fulbright-Nehru senior scholar award for research on climate change adaptation in India (2014–2016), and is a National Geographic Explorer grantee (2017–2018). Clouse holds a post-professional degree (SMArchS) in Architecture and Urbanism from the Massachusetts Institute of Technology and a BArch from the University of Oregon.

Routledge Research in Landscape and Environmental Design

Routledge Research in Landscape and Environmental Design is a series of academic monographs for scholars working in these disciplines and the overlaps between them. Building on Routledge's history of academic rigor and cutting-edge research, the series contributes to the rapidly expanding literature in all areas of landscape and environmental design.

Reimagining Industrial Sites
Changing Histories and Landscapes
Catherine Heatherington

Landscape Performance
Ian McHarg's Ecological Planning in The Woodlands, Texas
Bo Yang

Desert Paradises
Surveying the Landscapes of Dubai's Urban Model
Julian Bolleter

Ian McHarg and the Search for Ideal Order
Kathleen John-Alder

Walking, Landscape and Environment
David Borthwick, Pippa Marland and Anna Stenning

Climate-Adaptive Design in High Mountain Villages
Ladakh in Transition
Carey Clouse

For more information about this series, please visit: www.routledge.com/ Routledge-Research-in-Landscape-and-Environmental-Design/book-series/ RRLAND

Climate-Adaptive Design in High Mountain Villages

Ladakh in Transition

Carey Clouse

Routledge
Taylor & Francis Group

LONDON AND NEW YORK

First published 2021
by Routledge
2 Park Square, Milton Park, Abingdon, Oxon OX14 4RN

and by Routledge
52 Vanderbilt Avenue, New York, NY 10017

Routledge is an imprint of the Taylor & Francis Group, an informa business

British Library Cataloguing-in-Publication Data
A catalogue record for this book is available from the British Library

Library of Congress Cataloging-in-Publication Data
A catalog record has been requested for this book

ISBN: 978-0-367-42729-0 (hbk)
ISBN: 978-0-367-85462-1 (ebk)

Typeset in Sabon
by Wearset Ltd, Boldon, Tyne and Wear

Contents

Acknowledgments

The stories in this book come from Ladakhis, and their willingness to share their work, and their lives, has enabled this research. A special thanks to the staff at Leh Nutrition Project, LEDeG, and the Ice Stupa Project, whose organizational contact information can be found in the appendix. The author has also benefitted from countless conversations, collaborations, and conferences relating to Ladakh; some of the places where those events unfold are also listed in the appendix.

This work has been supported by a variety of organizations over the course of the past six years. Fieldwork was funded in part by a *National Geographic* Explorer's grant, a Fulbright-Nehru Senior Research Fellowship, and by numerous research grants received from the University of Massachusetts, Amherst.

Considering Ladakhi self-sufficiency under COVID-19

The long-term self-sufficiency of high-altitude subsistence farmers in Ladakh, is noteworthy, particularly in the context of the pressures of globalization. Today, this autonomy is, in part, due to the architecture, landscape architecture and infrastructural adaptations that have occurred under recent climate change challenges. However, the impact of the COVID-19 virus has also tested these measures of Ladakhi independence, and in telling ways. In considering even the earliest waves of the outbreak in Ladakh, the limitations of planned self-sufficiency come to light.

When the global pandemic struck travelers in January and February of 2020, Ladakh appeared to be a region that was well-insulated from global tourism, largely because very few travelers venture to the region during the winter months. Moreover, Ladakhi people typically experience reduced service roles in the winter months, working instead on farms, retreating to ancestral homes, and generally limiting contact with others.

The first COVID-19 cases identified in Ladakh arrived in March 2020, from a dozen residents who had returned from pilgrimages abroad. These travelers originated from Ladakh, but acquired the illness while abroad, thus bringing it into the region. The individuals were hospitalized while the rest of the region moved into quarantine. It appeared that this strategy worked: in late-March, Ladakh was declared COVID-free (Sharma 2020).

However, in the ensuing weeks, many hundreds of students returned from schools in other cities and states to shelter in Ladakh with their families. This caused an additional influx of travelers into the region, along with other residents returning home coming from across India and beyond. In April, additional cases were announced in local and national news sources, and in the weeks up to this writing, government news sites have continuously reported additional COVID-19 cases in both Leh and Kargil districts. Together, the two districts had a combined total of more than 1,000 positive cases by July 1, 2020 (*Tribune News Service* 2020).

In addition to the essential concerns about human health and welfare, Ladakh has also suffered a major economic loss, in the form of tourism, from the COVID-19 virus in 2020. Tourism is the central driver for Ladakh's contemporary economy, and without this income from 2020, repercussions will be felt in numerous sectors.

The first major impact will be seen in educational attainment. Ladakhi students pay tuition to attend many secondary schools and colleges, both regionally and nationally. Students whose families can't afford this tuition will need to pull their students from these high schools and colleges. This loss of education, professional preparation and the overall advancement of the Ladakhi workforce will likely have a lasting imprint on Ladakhi opportunities in the future.

A second impact has been felt in the realm of tourism, as entrepreneurs working in this sector will experience at least one season of lost income. Many individuals have taken out loans in order to build guest living quarters on their land, or to purchase vehicles or other business assets. These small-scale entrepreneurs could be in danger of losing their land, assets, and, ultimately, their means of employment if they cannot make payments on their loans this year.

If there is something to be learned from the case of Ladakh, remoteness, and the self-sufficiency that accompanies this geographic remove, can be a considerable benefit in the context of a global pandemic. Individual village households, as subsistence farming units, function well without many external inputs. Climate-adaptive design interventions all contribute to the functioning of households and farms, enabling ongoing production during quarantine.

Across the globe, the areas of impact of the virus on the economy, the environment, and society are far-reaching. But in Ladakh, the pandemic is also illustrative of gaps in the relative independence of a region that is both lauded for radical autonomy and also extraordinarily vulnerable to change. It is, then, perhaps useful to consider how climate change-adaptive design interventions have fallen short of assisting in ameliorating the problems of the pandemic, and also to acknowledge the region's capacity to pivot to greater resilience, given current COVID-19 challenges. Climate-adaptive design projects tend to impact and improve environmental, rather than human systems. Increasingly, the concerns of human, social, cultural, religious, political, and economic realms may be understood as embedded within, and central to, environmental functioning.

The position that many Ladakhis are in today, under a COVID-19 quarantine, is perhaps no different than the situation that is faced by individuals the world over. The future economic health of the region appears to be in a dire condition, pushed perhaps one way or another by the duration of the COVID-19 virus and its long-term impact on tourism. As in many other places, Ladakh has been split between the individuals who have little to lose (having no prior reliance on tourism for economic benefit or education, for instance), and those whose lives will be forever altered by the loss of income, or opportunity, that this pandemic produces.

References

Sharma, Arun. 2020. "Covid-Free Not Long Ago, Ladakh Sees 198 New Cases." *The Indian Express*, June 14, 2020. https://indianexpress.com/article/india/covid-free-not-long-ago-ladakh-sees-198-new-cases-6457813/.

Tribune News Service. 2020. "With 11 New Patients, Ladakh's Tally 1,002," July 4, 2020. www.tribuneindia.com/news/j-k/with-11-new-patients-ladakhs-tally-1-002-108196.

1 Introduction

From diminishing snowpack and glacial mass to the increasing frequency and severity of cloudbursts and glacial lake outburst floods, extreme environmental shifts have already begun to register in Ladakh, and in telling ways. Here climate change has become a new force for design in villages and towns, influencing new forms of architecture, infrastructure, and productive landscapes. For the rest of the world, such sites of transition serve as important bellwethers; as communities responding with regenerative and adaptive design interventions could ultimately help to inform future development patterns elsewhere.

During the course of the past three decades, dozens of subsistence agricultural villages in Ladakh have witnessed shrinking glaciers and an increasingly erratic supply of meltwater. Having relied on these relatively stable reserves for crop irrigation for centuries, today's farmers must now contend with a loss of water access, as well as irregular weather patterns and events, such as drought, mudslides, and cloudbursts. In the face of this shifting landscape, many farming villages have responded by altering longstanding agricultural practices in an effort to adapt to a changing climate.

The view from Ladakh

Villages across the Ladakh region, which lies between the Himalayan and Karakoran mountain ranges in north India, are experiencing increased water scarcity, primarily from diminishing snow pack and glacial mass caused by climate change (Mingle 2015; Rizvi 1998) (Figure 1.1). Subsistence farmers in this arid, mountainous region are almost wholly dependent on glacial meltwater to irrigate crops of barley, wheat, and vegetables. Lacking an adequate water supply for the foreseeable future, these ancient farming villages must each grapple with the same difficult decision: whether to move whole households closer to rivers, or to seek out new sources of water for their residents. In some cases, the Indian government has already intervened to build new irrigation canals, rerouting water across many miles of rough terrain (Grossman 2015). In other cases, the state and local government has offered to help relocate entire villages to new riverside sites. This displacement forces villagers to abandon locations of important cultural relevance and, in the case of Ladakh, nearly ten centuries of shared architectural, religious, and agricultural investment. For a number of these communities, changing the physical makeup of an existing village site is a more palatable prospect than resettling an entire group of households, and in these cases adaptive design projects have emerged. These user-driven design efforts suggest that in Ladakh and elsewhere, climate change offers an opportunity to reconsider the shape and form of architecture, infrastructure, engineering, and landscape architecture, as well as broader planning patterns.

Within the context of Ladakhi scholarship, the topics of architecture, landscape architecture, and urban design routinely reference the distant past rather than more current trends. In addition to connecting to the topics of climate change adaptation and resilience, this project fills an essential void in the design literature of an important but understudied region. In the design disciplines as well as in the national consciousness,

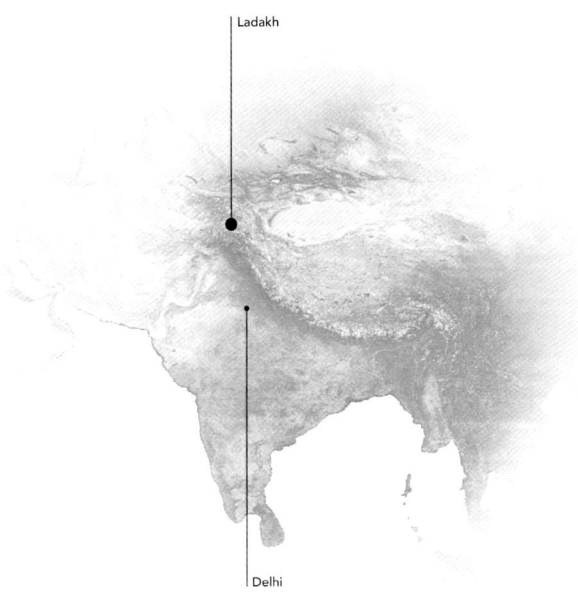

Figure 1.1 Map of India showing the location of Ladakh.

the topics of self-sufficiency and climate change have risen to the forefront of the collective consciousness. Whereas many high-tech global design solutions for climate change adaptation have already been addressed in literature and the media, this book highlights a different methodology and approach. For example, all of the climate-adaptive design techniques featured in this text involve affordable, accessible, regionally appropriate DIY approaches that are within reach of everyday farmers. Moreover, because many of the specific constraints of high-altitude mountain towns are similar across regions and countries, these designs suggest responses that could be instructive for other parts of the world.

This book addresses the wide-ranging design strategies and physical characteristics of a relatively new field of design inquiry: climate change adaptation. With an emphasis on nine of the most radical and creative design ideas currently employed in Ladakh, this research categorizes different thematic approaches to climate change design and simultaneously explains some of the myriad factors (social, cultural, religious, economic, environmental, and political) that have shaped this design thinking to date.

This book also fills an important gap in the literature on climate change adaptation in the design fields. As climate change increasingly becomes a topic that is studied in the design academy as well as in allied professions, new resources are needed, particularly those involving low-cost projects in a high-altitude context. The nine examples featured here help to create a realistic picture of the ways in which design thinking can have an impact on climate change adaptation, and the methods, concerns, and opportunities that this work involves.

Methods and structure

This book describes a series of nine climate-adaptive design strategies currently employed in northern India. These examples range from traditional design solutions to new prototypes, spanning scales and environmental constraints. Even so, four major themes surface repeatedly, as the topics of food security, water management, energy efficiency, and material scarcity have come to define climate change adaptation in Ladakh. Each of the different climate-adaptive design techniques reflect the region's unique and highly specialized response to climate stress. In each case study, the small-scale, community-based adaptation strategies practiced in Ladakh suggest a level of design thinking that supports a future of environmental stewardship, economic autonomy, cultural consciousness, and social cohesion.

With a focus on understanding design for climate resilience in Ladakh, this project integrates qualitative and interpretive research from the disciplines of architecture, landscape architecture, and urban planning. Design examples range in scale, scope, cost, and efficacy. In addition to farmers and communities acting on small, individualized projects, case studies include engineers building artificial glaciers high above existing villages, NGOs improving food security through greenhouse development, and monks who have invested in signature ice stupas. In sharing these examples together, the project offers a deeper look into the communitarian and extraordinarily resourceful region of Ladakh, while simultaneously providing ideas that could be deployed in other high-altitude resource-constrained communities.

Early chapters frame Ladakh's climate-adaptive design within the broader terrain of climate change science, design engagement, adaptation, and equity. Then, basic background information about Ladakh and the context of northern India help to bring readers up to date on historical information and recent research. These topics include weather, geography and isolation, contested borders, religion, family and village composition, and the role of outsiders. Ladakhis are living in a place that is widely recognized as one of the front lines of climate change, and while their specific geographical experience may be unique, their innovative responses could be instructive to designers elsewhere. To explore this broader relevance, the topics of climate, economy, development, and the legibility of the region are highlighted. This place-based background is then balanced with more universal global considerations, such as narratives of misery, state power, outsider engagement, and the topic of planetary urbanization.

A series of nine case studies follow, highlighting real-world, tested climate-adaptive design solutions. These examples include *Artificial Glaciers*: ice stockpiles at 13,000–15,000 feet; *Ice Stupas*: ice towers created to retain water (as ice) for use in the spring; *Snow Barrier Bands*: stone walls used to catch and divert snowfall at 17,000 feet; *Solar Design*: passive solar design, photovoltaic electricity, and solar hot water; *Greenhouses*: year-round agriculture; *Reservoirs and Canals*: water management and dispersal; *Tree Planting and Tree Armor*: material production and ecosystem services; *Food Security*: the Indian ration program and other collaborative forms of assistance; and *Recruiting Allies*: forging relationships for development work.

These examples ground a discussion of larger, cross-cutting themes, which represent more universal constraints under the pressures of climate change, and may be of use to designers grappling with similar issues in other contexts. Topics such as cultivating support from external allies, and the inclusion of the design disciplines with regard to climate-adaptive design and development must be addressed in both opportunistic and critical frames. Topics such as sponsorship and aid; NGO involvement; research in the region; government support; and fundraising help to demonstrate the relevance that this content could have for other parts of the world. The book concludes by comparing the forced nature of these Ladakhi examples of design adaptation to a more intentional path toward sustainability at the global scale.

The role of design

In the face of global climate change, planners and designers can provide valuable expertise to communities in transition. Indeed, changing environmental conditions will necessitate new solutions for the mitigation and adaptation of infrastructural systems, changing human settlement patterns and traditional ways of conceptualizing the built environment. While the effects of a changing climate are difficult to project and plan for, instability itself will likely be the chief characteristic of future development. In this environment, resilient, adaptive, and flexible designs for supporting human development suggest a way forward.

The design community (made up of, among others, architects, landscape architects, urban designers, and planners) has both the expertise and the responsibility to engage with communities confronting climate change. This book illustrates both the power of design agency in tackling environmental problems, and the importance of inviting designers to participate in such palliative efforts. As the volatile environmental conditions of the twenty-first century begin to impact global cities and towns, the broad reach and disciplinary expertise of architecture, landscape architecture and urban planning could provide valuable design thinking around climate change geoengineering, adaptation, and mitigation.

After all, the complicated problems of climate change adaptation and design for resilience demand holistic design thinking, best made by groups of people with varied experience and expertise. Designers have too often been left out of conversations around climate change adaptation, despite their problem-solving skills, visioning expertise, and ability to effectively engage multiple stakeholders. And yet, the inclusion of designers in the decision-making processes around climate change challenges is not enough. Stakeholders also need to be brought around the table, working with designers to make connections between formal and informal modes of knowledge and wisdom.

In an effort to find inroads to broader dialogue, and opportunities for a more expansive practice, non-traditional modes of practice must be linked to the staid traditions of the design disciplines. In referencing vernacular, DIY, or outsider models for self-sufficiency and design thinking, this book illustrates the power of bootstrapping design efforts in tackling pressing environmental problems. But designers routinely view these types of interventions as outside their professional ambit, and thus may miss out on the important lessons that they provide. Instead, designers might learn from the explorations that are underway at the margins of their disciplinary range, and in so doing, bring those ideas closer to home. After all, environmental crises present opportunities for designers to engage in valuable community service through their work, while simultaneously helping to expand the range and relevance of the design disciplines.

Just as the effects of climate change and resource scarcity are becoming visible across the globe, a host of emergent creative design responses also follow. The Anthropocene has issued in a challenging new frontier for human settlement, one that demands creative design thinking around adaptation, requires greater levels of resilience built into systems, and necessitates equitable development. Ladakhi design adaptation offers a view from one small outpost of civilization – a glaciated mountain region already hard hit by climate change – where compelling new methods of discovery, innovation and resilience stand on stark display.

References

Grossman, Daniel. 2015. "As Himalayan Glaciers Melt, Two Towns Face the Fallout." Yale E360. March 24, 2015. https://e360.yale.edu/features/as_himalayan_glaciers_melt_two_towns_face_the_fallout.

Mingle, Jonathan. 2015. *Fire and Ice: Soot, Solidarity, and Survival on the Roof of the World*. 1st edition. New York: St. Martin's Press.

Rizvi, Janet. 1998. *Ladakh: Crossroads of High Asia*. Delhi; New York: Oxford University Press.

2 Background on Ladakh

It is difficult to comprehend the climate-adaptive design responses underway in Ladakh without first gaining a solid understanding of the regional context. This is probably true for most locations, but especially so in Ladakh, where "Technology developed for other parts of the world appears to have little relevance to this region of peculiar climate and soil conditions" (Prakash 1991, 27). Ladakh is exceptional in many ways, with specific, tailored environmental, social, cultural religious, political, and economic frameworks. To study climate change adaptation in Ladakh, then, is to first learn about how the population has adapted to fit the region.

Climate and context

Ladakh is a mountainous region, located between the Karakoram and Great Himalayan ranges in north India (Figure 2.1). Here high-altitude glaciers and snowfields feed scant alpine streams, and these rivulets combine to form the headwaters of the Indus River. Despite visible surface water and snowy passes, this is predominantly a rugged and dry landscape, referred to as a high-altitude desert (Akhtar and Gondhalekar 2013, 25; Daultrey and Gergan 2011) or a semi-arid arctic climate (Demenge 2013). Indeed, 68 percent of the land area is 16,000 feet or more above sea level, and the region annually receives nearly 300 days of sun, on average (Prakash 1991). Temperatures fluctuate between −30 degrees Celsius (−22 °F) in the winter and 25 degrees Celsius (77 °F) in the summer, and scarce rainfall, between 50 and 300 mm annually, has historically defined the region's climate (Daultrey and Gergan 2011; Demenge 2007; Nüsser, Schmidt, and Dame 2012a). To many visitors, Ladakh's environmental conditions appear to be both extreme and inhospitable, characterized by high altitude, low humidity, and extraordinarily low precipitation.

Because of these environmental constraints, Ladakh's mountainous terrain is sparsely populated, containing some of the highest inhabited villages on earth (Prakash 1991). The majority of the more than 50 villages that exist in this area are several hundred years old and support small, self-governing populations of between 100 and 1,500 people (District

Figure 2.1 A snowy pass high above Ladakh.

Statistics & Evaluation Office 2013). Unlike the cities and towns in northern India, villages typically practice a single integrated system of subsistence agriculture and have limited opportunities for other types of economic participation. Despite the enduring sustainable human development in this region, water scarcity and attendant environmental pressures have historically curbed growth. Today, these existing environmental pressures, coupled with unstable weather patterns caused by climate change, have started to force climate adaptation measures.

In this challenging environment, strong social and cultural traditions have effectively tethered people to the land, and to each other. According to scholar R.S. Mann, Ladakhi "people feel that their adaptation (to climate) alone made them survive whenever nature posed (a) threat to their existence" (Mann 1986, vi). While this viewpoint alone cannot adequately encompass the range of people's experiences today, the harsh realities of the weather and geography in Ladakh still fundamentally impact the region's social, cultural, and economic characteristics.

Ladakh is made up of two distinct districts: Leh District and Kargil District, although this book focuses primarily on Leh District (Figure 2.2). Ladakh's Leh District provides a more uniform compositional and climactic region for study, and so in this analysis, the title 'Ladakh' serves as a sort of shorthand for the Leh District region, and the two terms are used interchangeably. Some 147,104 people live in Leh District, with a population density of roughly three people per square kilometer (District Statistics & Evaluation Office 2013). The region is composed of predominately

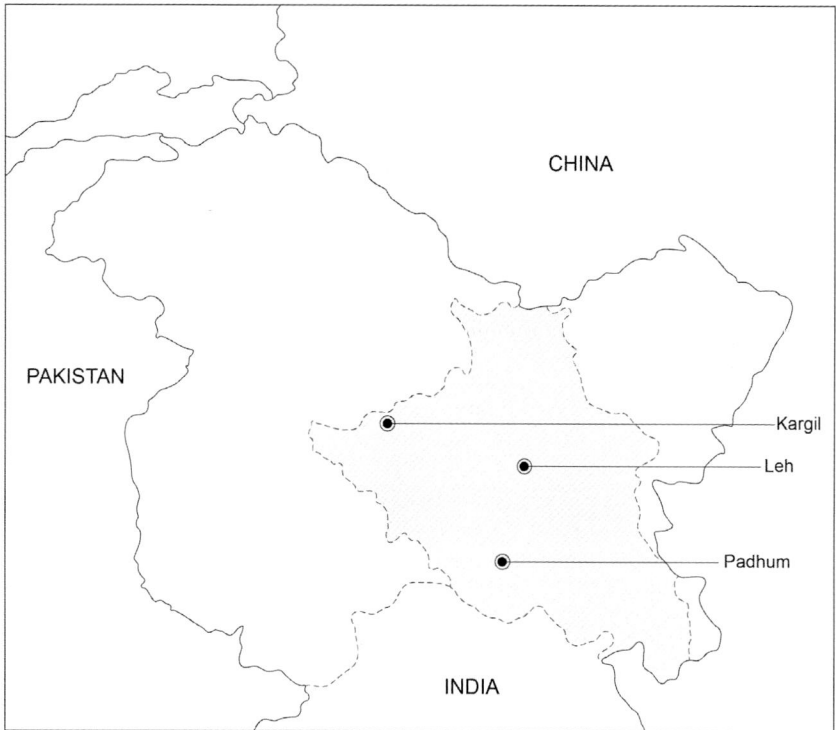

Figure 2.2 Map of Ladakh.

Buddhist and Muslim households, with a small number of other religious groups, including Christian and Hindu sects. Similarly, a constellation of languages persist in the region: Ladakhi, Tibetan, Balti, and Urdu are the official languages, while Hindi and English are common secondary spoken languages.

Agriculture in Ladakh

While the majority of the land in this area is too high, too dry, or too steep to farm, most Ladakhi villages have been opportunistically sited in the folds of mountains where they capture glacial meltwater as it flows downstream (Figure 2.3). By redirecting this water into channels and reservoirs, farmers irrigate fields, build soil fertility, and grow high-altitude, cold-hardy crops. Agricultural outputs are limited by a combination of factors, including high altitude, low humidity, and extremely low precipitation (Angchok and Singh 2006). Villages across Ladakh tend to practice the same integrated system of subsistence agriculture, with surprisingly high

Figure 2.3 Villages are striking green seams in the desert landscapes.

yields that have been credited with supporting widespread prosperity and social cohesion (Crook and Osmaston 1994, 140). Despite recent demographic shifts and new alternative opportunities for economic participation, this agricultural effort, combined with animal husbandry, forms the backbone of village life in Ladakh (Nüsser, Schmidt, and Dame 2012b) (Figure 2.4).

The primary crops grown in Ladakh are barley, peas, wheat, buckwheat, millet, mustard seed, apricots, and livestock fodder, including alfalfa (Gutschow 1993; Humbert-Droz and Dawa 2004). Apple and walnut trees can thrive in low-elevation areas. Small kitchen gardens, and larger plots of carrots, turnips, potatoes, and lettuces may be planted around village households (Nüsser, Schmidt, and Dame 2012b). While this agricultural production continues to support subsistence households, recently some farmers have shifted land holdings to the production of market crops, in response to the growing presence of the Indian military and tourist industry and their demand for potatoes and vegetables (Tiwari and Gupta 2007). The primary market for the region is agriculture, animal husbandry, and trade, although tourism is the region's main economic driver.

While the region appears to be relatively isolated today, Ladakh has witnessed many sweeping changes over the last five decades, particularly in

Figure 2.4 Grazing supports agriculture in Ladakh.

the rise of tourism and in an increasing number of employment opportunities in the local military and government sectors. According to scholar Mohammed Deen Darokhan, this shift has also caused a deterioration in the practice of ancient farming techniques and in overall agricultural reliance. Darokhan regards agriculture as "the only sustainable way of life in Ladakh," and suggests that villages could "lose the genetic material and cultivation techniques due to a neglect of farming" (Darokhan 1999, 79). Other scholars note that in Ladakh, agriculture is inextricably tied to land stewardship and cultural survival, as the traditional agricultural lifestyle has supported many centuries of sustainable human settlement in the region (Norberg-Hodge 1991; Rizvi 1998).

While these demographic shifts have become even more marked since the Indian government opened Ladakh to tourists in 1974, traditional agrarian values persist in cultural, social, and even economic spheres (Dollfus and Labbal 2009). Many Ladakhi village households still actively farm their historic land holdings and have few opportunities for alternative economic participation. This is the outcome of many years of farming effort and collective decision-making, as generations of farmers have worked with natural cycles "to green the desert and bring biological diversity" to an otherwise barren landscape (Darokhan 1999, 78). Indeed, the lush green farmlands, thick stands of poplar and willow trees, and surprisingly fertile topsoil visible in an otherwise inhospitable desert environment represent many centuries of judicious land stewardship. Perhaps not surprisingly, this difficult agricultural environment also reinforces the value of traditional methods, where "the adoption of agricultural innovations is slow" (Mann 1986, 173). Under the pressures of a changing climate, agricultural adjustments are both needed and born out, and scholars Nüsser, Schmidt, and Dame argue that in Ladakh this adaptation happens in conjunction with a reliance on older methods (Nüsser, Schmidt, and Dame 2012b).

Water

Historically, precipitation in the form of rain has been so minimal that most of the water used for farming comes instead from melting snowfields and glaciers in the northern part of the watershed (Akhtar and Gondhalekar 2013). In most villages, irrigation potential is limited by the natural capacity of the watershed located above them, and ancient hand-built channels capture meltwater for judicious use by farmers below. Below these villages, on the valley floor, the headwaters of the Indus River takes shape, growing from a series of tiny streams into a tumultuous river body. Across Leh District, much of the river's edge lacks "viable arable land" and, anyway, the prospect of pumping water from rivers below is, in this context, too difficult.[1] Instead, villages and farmlands drape across the sides of mountains, sited between the high glaciers above and the river below.

While the presence or absence of glacial meltwater may have initially determined the location of a village, the resulting system of irrigation infrastructure depicts a far more intentional, sophisticated and laborious planning effort. Irrigation is a prerequisite for the production of vegetables in Ladakh and indeed this practice has emerged as a means of contending with an otherwise inhospitable climate and landscape. According to Tibetan scholars, skillful irrigation was present in this region in the tenth century, and in many villages early stone canal work has been improved upon for centuries (Bell 1928).

In this climate and context, irrigation has grown into an elegant, efficient, and equitable system in most villages, honed over many hundreds of years and kept in check by village politics and social norms. Scholar Kim Gutschow has described the power dynamic created by this water management as "webs of water," which bind villagers to each other and to a communal decision-making process (Gutschow 1993). Sharing a single water source has led stakeholders to develop an advanced rotational system for water distribution as well as enduring practices for collective bargaining and oversight. Often the fields are assigned to a rotation of night or daytime watering, so that every hour, water is rationed and collected. For those fields assigned to a night watering schedule, the farmers might divert the water to a storage tank or *zing* to hold it for use the following day. In the summer months, fields might be watered every seven to ten days, depending on the crop and time of the season (Gutschow 1993). Because meltwater is a shared resource, equitably divided across an array of field parcels located at different elevations, collective oversight and investment characterize village farming practices. If water ceases to flow as glacial reserves dry up in the summer months, all of the farms across the full extent of the village downstream will be impacted by the drought.

In Ladakhi, the annually elected managers of the water are called *chu dpon*, or "Lords of the Water" (Gutschow 1993). They serve the entire village during their term by splitting up into teams to check that the water has been shut off at various channels, monitoring assigned channels and rotating in teams of two to sleep by storage tanks to ensure that collected night water gets allocated to the proper fields the next morning (Gutschow 1993). Gutschow notes that these individuals, along with a more communitarian surveillance and village affiliation, keep farmers from taking more than their entitled shared of water. This is an ancient farming practice that relies almost entirely on the relative stability of village farming holdings and the regularity of seasonal environmental conditions. As town populations grow, weather patterns shift, or as people in villages desire more independence and autonomy, this system could easily fall apart (Gutschow 1993, 114).

In Ladakh's Leh District, water is primarily used for two purposes: household activities and crop irrigation. Most villages supply household water from groundwater, rather than the surface meltwater used in agriculture. In these cases, groundwater is drawn from wells (using hand pumps), mountain springs with dedicated "clean" taps, or in more heavily populated areas from those two sources via tanker trucks, which refill household barrels or tube wells (Akhtar and Gondhalekar 2013, 25; Sudhalkar 2010) (Figure 2.5). While the challenges associated with groundwater depletion in this region are significant, household water use represents but a tiny fraction of the total water volume needed to support farming and animal husbandry. As such, surface water management – the collection, retention, and dispersion of flowing meltwater from glaciers and snowfields above – has become a central planning concern of the Ladakhi village (Daultrey and Gergan 2011, 4–5).[2] Moreover, subsistence agricultural practices, and the elaborate system of irrigation associated with these efforts, dominate the social, economic, and physical structures of village life.

According to Dawa et al., agriculture draws upon about 230 different watersheds in Ladakh, and these watersheds are fed by a combination of rain, snowmelt, glacier melt, springs, and marshes (Dawa, Dana, and Namgyal 2000, 235). Most villages in the region of Zanskar are much smaller than those in the rest of Ladakh, so the irrigation rotation can sometimes function at the scale of the household rather than the water channel (Gutschow 1993, 113). This is a far more flexible system, but limited in terms of geographic spread.

Figure 2.5 A springline transports potable water from high elevation down to a village for use.

Contested borders

Because Ladakh occupies a strategic geopolitical location, and has ongoing border disputes with China and Pakistan, Ladakh has hosted a year-round military presence since 1952. Ladakh officially became a Union Territory in 2019, but was previously designated a semi-autonomous region in the state of Jammu and Kargil, India's northernmost territory. This region occupies two contested borders: one shared with China in Saichen and one shared with Pakistan along the Nubra border and along the line of control near Kargil. Since India's independence, then, the character of Ladakh has transitioned from a major Trans-Himalayan trading hub to a site of geopolitical significance, as a borderlands region.[3]

Leh District shares contested borders with both Tibet and Pakistan, and despite the relative rural context, hosts an active and extraordinarily visible Indian military presence (Figure 2.6). India's interest in a secure and stable border ensures a strong military presence in the region, and the Indian military has invested in roads, bases, and buildings that support troops. Beyond this built infrastructure, the soldiers stationed throughout the region represent an enormous economic opportunity for local farmers and entrepreneurs, as the number of people in the military in this area are nearly as populous as the civilian population (Daultrey and Gergan 2011).

Ladakh's position at the junction of two contested borders also provides a powerful reason for the Indian government to support the economic development and modernization of the region. While Ladakh lacks a population that could significantly sway the economy or policy of India, it occupies an extremely sensitive geopolitical location, and one that is constantly at risk of defection. In an effort to support and advance the region, the Indian government has sponsored enormous infrastructural projects such as irrigation canals and hydropower dams, the development of hospitals, schools and government offices, and an infusion of capital to improve roads in the area.

Religious context

While much of the Indian subcontinent can be understood in terms of major Muslim, Hindu, and Christian influences, Ladakh draws much of its religious and cultural context from the ideologies shared by adjacent nations. The region's predominant Buddhist and Muslim makeup reflects an enduring association with the nearby populations of Pakistan and Chinese-controlled Tibet. These two major religions – Buddhism and Islam – have influenced and at turns dominated the area, although today these two groups have settled together in Ladakh. These two major religions, and other smaller sects as well, are visible in the historic temples and mosques, stone sculptures, and other holy spaces that dot the landscape. Shared religious space is a constantly unfolding dynamic that includes real friction within towns and

Figure 2.6 Indian troops walk on the road outside of Leh.

neighborhoods due to property and resource disputes, as well as many examples of peaceful coexistence. This religious complexity, however, challenges the idea of a timeless and peaceful homogeneous Ladakh, the image most often sold by the tourism industry (Aggarwal 1997).

The visibility of Tibetan refugees in the area, as well as regular visits from His Holiness the Dalai Lama and the influx of Buddhist tourism during the course of the past five decades, has led to the popular characterization of Leh District as a Buddhist territory. Indeed, foreign scholars have referred to Ladakh as a "Little Tibet" (Crook and Osmaston 1994, 140), a label rejected by contemporary scholars such as Quentin Devers, who note that the area "includes a much richer set of material, cultural and religious influences" (Devers 2018, 103). This characterization, like the quaint village idealism promoted by Ladakh's tourism industry, is not just incorrect but also profoundly limiting. The region, like all of India, is a dynamic living place, made up of a complex assortment of different factors, each of which is capable of change over time.

Today it is more common for residents, scholars, and activists to push back against outdated and romantic notions of Ladakh, challenging visitors and government agencies to recognize the region for all its dimensions. Even so, misconceptions still obscure literature on Ladakh, perhaps even with more pervasiveness than those representations produced by foreign travelers. Scholar Ravina Aggarwal notes that "If Western academia has glossed over contemporary political realities, tending to dwell in wistful moments of paradise lost, Indian academics have also contributed to the process of partial representation" (Aggarwal 1997, 25).

Economy

Despite fundamental environmental and development limitations, Ladakh can hardly be called an underdeveloped society. Villages still showcase the lasting imprint of many traditional lifestyle and farming practices. However, within this long-established agrarian landscape, new modes of economic activity are also evident. The town of Leh has become the central hub for the region's burgeoning tourism industry, which discharges thousands of vehicles – buses, motorcycles, taxis, jeeps, and private automobiles – out to the treks, passes, lakes, and other tourist enclaves in the region every summer. Tourism has created economic opportunities at every level; directly for porters, guides, hoteliers, homestay hosts, and retailers, and, indirectly, for all of the associated producers that enable this market to thrive.

In addition to the tourism industry, the militarization of this border region has become a significant economic driver in both direct and indirect ways. Roads, canals, dams and other major infrastructural projects in the area were financed or otherwise supported by the national army, creating jobs and material needs. In addition to these development projects, the army buys food, goods, and fuel in this region.

Isolation and self sufficiency

While Ladakh is geographically remote, the region is not isolated per se; residents maintain strong local, national and global connections through trade relationships and social ties. The relative isolation that characterizes the region and its people is, perhaps, ultimately a recent construct, fueled by the somewhat detached perspectives of the Indian government and foreign visitors. A number of scholars have argued, instead, that Ladakh has long been a crossroads of trade, and could even be considered to have cosmopolitan processes (Fewkes 2011). Today the area remains connected to the rest of the world via two major roads, which are open for about seven to eight months of the year, and a major airport, open year-round, in the town of Leh.[4]

Historically, the ancient Silk Roads cast a lasting imprint on Ladakh, by solidifying trade relationships and imparting new cultural, physical, and economic influences. Other trade routes, which strengthened or slackened according to political and economic conditions over the centuries, also wound through the region and connected Ladakh to Kashmir, Baltistan, India, Lhasa, and Tarim Basin (Devers 2017). Salt and pashmina were always exported via Ladakh, but especially as gold and other resources declined in the 1700s, trade routes become more limited (Devers 2018). Although Ladakh was a veritable hub for national and international trade prior to the second half of the twentieth century, the closure of the border by the Chinese in 1949, and ensuing friction with Pakistan effectively blocked these routes in the following decades (Rizvi 1998). After partition, Ladakh became an inaccessible destination even for Indian tourism, and only re-opening to national and international tourists in 1974 (Akhtar and Gondhalekar 2013).

When India closed its borders with Pakistan and Tibet in the mid-twentieth century, Ladakh became a frontier territory. Scholar Jaqueline Fewkes suggests that the vibrant trading history of the past continues to inform an understanding of Ladakhi identity, "as the roots of contemporary social, political, and economic contexts" (Fewkes 2009, 162). Despite this identity, the physical barrier of the trans-Himalayan mountain range has also undoubtedly cut the region off from the southern portion of the subcontinent, where the "forbidding climate, remoteness and inaccessibility (has) kept Ladakh isolated, except for traders, for centuries" (Mann 1986, 3). It is this juxtaposition, then, that has come to define Ladakh – as a region at once connected by trade lines and also separated by geography (Devers 2017).

The region, as documented by scholars John Crook and Henry Ostmanson in the late 1970s and early 1980s, appeared to be largely unchanged from conditions documented in earlier centuries. At this time, the single road into the region was difficult to travel and only legally opened to outsiders in 1974. Thus, this area was protected – by both climate and topography – from the changes and impacts of the surrounding regions. While most economic and political opportunities appear in more well-connected areas, scholar James C. Scott has identified a series of geospatial benefits associated with this kind of isolation. Scott suggests that mountain people can avoid tangible burdens of state, such as taxation and conscription, but also achieve intangible benefits as well, such as a more robust group identity, more day to day autonomy, and a relative remove from the power struggles that occur in a more connected state (Scott 2009, 9). Ladakh has always been a borderlands region, occupying a location that in and of itself may be considered protective.

Indeed, Ladakh's geopolitical remove has effectively created a largely self-sufficient and self-reliant system of villages. Because of the harsh climate, traditional ways of life – from family structure to land use to the types of crops produced – reflect a carefully considered contextual response. Moreover, if any one of these variables changes, it can cause a ripple effect on other parts of the system. For example, the three major crops prevalent in Ladakh suggest both the relative lack of agricultural diversity of the system, and the hyper-specificity needed by niche plants. Likewise, family structure was once organized around the structures of religious celibacy and polyandry to ensure stable population growth over time. These techniques gave rise to high-altitude subsistence farming regimes, finely tuned to the land and climate. In a more general sense, the practice of resource conservation allowed Ladakhi people to have healthy and productive lives despite minimal water, soil, energy, and material resources.

Central to the longevity of Ladakhi culture has been the value that village communities place on environmental adaptation and attendant self-sufficiency. One striking example of this adaptation is visible in the finely tuned irrigation and land management practices that underpin the region's high-altitude subsistence agriculture (Daultrey and Gergan 2011). Although Ladakhi villages are characterized by low soil fertility and a short growing season, farmers have historically produced excellent grain yields using high-altitude varieties of barley, wheat, and peas (Crook and Osmaston 1994). This ability to capitalize upon, and effectively manipulate, scarce natural resources has led researchers and visitors to extol the region as a model for self-sufficiency and stewardship in the face of extreme environmental conditions.

The rich cultural and social traditions exhibited in this area draw upon many centuries of lived experience: Ladakhis have developed sophisticated techniques to sustainably support themselves at altitudes above 11,000 feet (Darokhan 1999; Mingle 2015). In subsistence agricultural practices, medicine, education, social welfare, religious life, and even vernacular building techniques, the Ladakhi people have built a society that is independent, radically communitarian, and overwhelmingly sustainable (Norberg-Hodge 1991). The region's geospatial isolation has until recently demanded that villages remain relatively self-sufficient, and as a result, the Ladakhi people have developed elaborate place-based farming practices and buildings that support this lifestyle.

Current trends

Today, both human and natural systems in Ladakh are undergoing striking observable change (Daultrey and Gergan 2011). Here traditional and contemporary practices are not two separate poles – old and new – but instead appear to be a dynamic fusion of ancient practices and modern conveniences (Tiwari and Gupta 2007). While subsistence agriculture once dominated Ladakhi village life, new economic and social opportunities have opened in local governance, the military sector, and through tourism. An increasing population, including large numbers of visitors and immigrants, have helped to fuel a transition away from subsistence agriculture and help to paint a picture of a different, more diverse future. According to scholars Tiwari and Gupta, "These changes in the livelihood options and in the agrarian system have affected land use patterns in the town of Leh," (Tiwari and Gupta 2007, 218) and it is reasonable to assume that these trends will continue to move out to smaller villages across the region over time.

Indeed, Leh town and the nearby Tibetan refugee settlement of Choglamsar have witnessed explosive growth due to an influx in immigrants, boarding school students, and migrants from villages seeking new avenues for work. Today the typical central Ladakhi household is relying upon a combination of subsistence farming practices and off-farm opportunities for their livelihoods (Daultrey and Gergan 2011). This demographic shift marks an increase in available employment in cities and towns, and associated labor shortages for agricultural efforts in small villages.

In addition to these demographic shifts, the overall population of Leh District has significantly grown in recent years.[5] Scholar Jonathan Demenge notes that agriculture in this region has sustained a visible and clearly defined ecological footprint, one in which the biocapacity of the land has been carefully tuned over centuries, through irrigation, fertilization, and land husbandry. However, even within the bounds of this limiting ecological scope, Demenge suggests that "population growth and increasing human needs may lead to a better management of natural resources, agricultural extensification and intensification, and therefore to an increase in the total biologically productive area in the region" (Demenge 2007, 19). Demenge joins a group of scholars who anticipate increased use of Ladakh's resources over time, and also call for improved efficiencies to manage these additional loads.

At sites that draw tourists or host a military presence, village land holdings also shapeshift. Agricultural space has, across Ladakh in recent years, transformed to accommodate new economic opportunities, such as guest-houses, restaurants and retail space. According to Tiwari and Gupta,

> Over the years, as the population has increased and land fragmentation has begun to take place, houses have spread and come up in the form of hotels, restaurants and guesthouses, leading to the conversion of agricultural plots to other uses.
>
> (Tiwari and Gupta 2007, 218)

The large town of Leh characterizes this land transition, and acts as a central hub for tourism. According to the state tourism department, some 49,477 international tourists and 327,366 domestic tourists visited Leh in 2018 (India 2019).

Ladakhi village life has long been sustained through the careful management of resources, as in a closed loop system. The addition of new residents, visitors, and military troops may alter this balance with competing resource needs. While the nature of life in Ladakh has always been marked by change, recent geopolitical opportunities and events have also served to speed up this process. As early as the 1980s, Crook and Osmaston noted that the Ladakhi cultural and physical landscape was in transition. At that time, they cautioned:

> With such a closely integrated system, any social or technical changes may have unforeseen effects, and well-meant ideas to improve it need skeptical consideration. Some changes are probably desirable and inevitable, such as the increase in opportunities for independent and non-celibate careers for younger brothers and for sisters, resulting in a rise in population which is already evident.
>
> (Crook and Osmaston 1994, 192)

Today, the changes impacting Ladakh reflect many of these ongoing trends as well as a new giant that eclipses the region's religious, social, cultural, political and economic concerns and may well impede its progress: the environment in the region, once harsh but relatively stable, has become far more erratic under climate change.

Notes

1 This would require more energy and infrastructure, with attendant government support (Gutschow and Mankelow 2001, 28).
2 According to Daultrey and Gergan, "The main source of irrigation in Ladakh is surface area, with approximately 10,190 hectares of land around the tributaries to the Indus irrigated by a sophisticated and carefully managed system of small, hand-built mud canals, which make effective use of seasonal run-off from melting snow and ice at high altitudes. Irrigation using groundwater is negligible" (Daultrey and Gergan 2011, 4–5).

3 Since the India-China war in 1962, the vibrant trade between Tibet and Ladakh has effectively been stopped (Dawa, Dana, and Namgyal 2000, 238).
4 The Leh-Manali and Leh Srinigar roads are typically open from June to October, depending on the annual snowfall that closes high mountain passes.
5 The 2011 Census noted that Leh District grew from 117,232 inhabitants in 2001 to 133,487 a decade later (Census of India 2011).

References

Aggarwal, Ravina. 1997. "From Utopia to Heterotopia: Towards an Anthropology of Ladakh." In *Recent Research on Ladakh 6: Proceedings of the 6th International Colloquium on Ladakh, Leh, 1993*, edited by Henry Osmaston and Tsering Nawang, First Asian Edition, 21–8. Delhi: Motilal Banarsidass Publishers.

Akhtar, Adris, and Daphne Gondhalekar. 2013. "Impacts of Tourism on Water Resources in Leh Town." Vol. 30. Leh, Ladakh: International Association for Ladakh Studies.

Angchok, Dorjey, and Premlata Singh. 2006. "Traditional Irrigation and Water Distribution System in Ladakh." *Indian Journal of Traditional Knowledge* 5 (3): 397–402.

Bell, C. 1928. *The People of Tibet*. Oxford: Clarendon Press.

Census of India. 2011. "District Census Handbook: Leh." Part XII-A. 02. Jammu & Kashmir: Directorate of Census Operations.

Crook, John, and Henry Osmaston. 1994. *Himalayan Buddhist Villages: Environment, Resources, Society and Religious Life in Zangskar, Ladakh*. Bristol: University of Bristol.

Darokhan, Mohammad Deen. 1999. "The Development of Ecological Agriculture in Ladakh and Strategies for Sustainable Development." In *Ladakh: Culture, History, and Development between Himalaya and Karakoam*, edited by Martijn van Beek, Kristoffer Brix Bertelsen, and Poul Pedersen, 78–91. Denmark: Sterling.

Daultrey, Sally, and Reuben Gergan. 2011. "Living With Change: Adaptation and Innovation in Ladakh." *Climate Adaptation*. Available: www.yumpu.com/en/document/view/25089047/living-with-change-adaptation-and-innovation-in-our-planet.

Dawa, S., D. Dana, and P. Namgyal. 2000. "Water Harvesting Technologies and Management System in a Micro-Watershed in Ladakh, India." In *Waters of Life-Perspectives of Water Harvesting in the Hindu Kush-Himalayas: Volume II*, edited by SR Chalise and M Banskota, 235–59. ICIMOD. http://lib.icimod.org/record/22407.

Demenge, Jonathan. 2007. "Measuring Ecological Footprints of Subsistence Farmers in Ladakh." Cardiff, Wales UK: International Ecological Footprint Conference. http://web.mnstate.edu/robertsb/307/ANTH%20307/ecologicalfootprintfarmersladakh.PDF.

Demenge, Jonathan. 2013. "The Road to Lingshed: Manufactured Isolation and Experienced Mobility in Ladakh." *HIMALAYA, the Journal of the Association for Nepal and Himalayan Studies* 32 (1): 51–60.

Devers, Quentin. 2017. "Charting Ancient Routes in Ladakh: An Archaeological Documentation." In *Interaction in the Himalayas and Central Asia: Processes of Transfer, Translation and Transformation in Art, Archaeology, Religion and Polity*, edited by Eva Allinger, Frantz Grenet, Christian Jahoda, Maria-Katharina Lang, and Anne Vergati, 321–38. Vienna: Austrian Academy of Sciences Press. www.academia.edu/39101653/Charting_Ancient_Routes_in_Ladakh_An_Archaeological_Documentation.

Devers, Quentin. 2018. "Archaeological Ladakh: Recent Discoveries Redefining the History of a Key Region between the Pamirs and the Himalayas." *Central Asiatic Journal* 61 (1): 103–32.

District Statistics & Evaluation Office. 2013. *Blockwise Village Amenity Directory, 2012–13, Series 14*. Leh, Ladakh: District Statistics & Evaluation Office.

Dollfus, Pascale, and Valérie Labbal. 2009. "Ladakhi Landscape Units." In *Himalayan Landscapes Over Time: Environmental Perception Knowledge and Practice in Nepal and Ladakh*, edited by Joëlle Smadja, 85–106. Pondicherry: Institut Francais de Pondichéry. www.abebooks.com/Reading-Himalayan-Landscapes-Over-Time-Environmental/1386152000/bd.

Fewkes, Jacqueline. 2009. *Trade and Contemporary Society along the Silk Road: An Ethno-History of Ladakh*. New York: Routledge. https://academic.oup.com/ahr/article/115/2/513/11286.

Fewkes, Jacqueline. 2011. "Living in the Material World: Cosmopolitanism and Trade in Early Twentieth Century Ladakh." *Modern Asian Studies*. https://doi.org/10.017/S0026479X07002058.

Gutschow, Kim. 1993. "Lords of the Fort, Lords of the Earth, and No Lords at All: Politics of Irrigation in Three Tibetan Societies." In *Recent Research on Ladakh 6: Proceedings of the 6th International Colloquium on Ladakh, Leh, 1993*, edited by Henry Osmaston and Tsering Nawang, 105–15. Leh, Ladakh.

Gutschow, Kim, and Seb Mankelow. 2001. "Dry Winters, Dry Summers: Water Shortages in Zangskar." *International Association for Ladakh Studies* 15 (August): 28–32.

Humbert-Droz, Blaise, and Sonam Dawa. 2004. *Biodiversity of Ladakh: Strategy and Action Plan*. Leh, Ladakh: LEDeG.

India, Press Trust of. 2019. "Leh Emerging Favourite Destination for Foreign Tourists, Nearly 50k Visited in 2018." *Business Standard India*, January 19, 2019. www.business-standard.com/article/pti-stories/leh-emerging-favourite-destination-for-foreign-tourists-nearly-50k-visited-in-2018-119011900373_1.html.

Mann, Rann Singh. 1986. *The Ladakhi: A Study in Ethnography and Change*. Delhi: Anthropological Survey of India.

Mingle, Jonathan. 2015. *Fire and Ice: Soot, Solidarity, and Survival on the Roof of the World*. 1st edition. St. Martin's Press.

Norberg-Hodge, Helena. 1991. *Ancient Futures: Learning from Ladakh*. Sierra Club Books.

Nüsser, Marcus, Susanne Schmidt, and Juliane Dame. 2012a. "Irrigation and Development in the Upper Indus Basin." *Mountain Research & Development* 32 (1): 51–61. https://doi.org/10.1659/MRD-JOURNAL-D-11-00091.1.

Nüsser, Marcus, Susanne Schmidt, and Juliane Dame. 2012b. "Irrigation and Development in the Upper Indus Basin: Characteristics and Recent Changes of a Socio-Hydrological System in Central Ladakh, India." *Mountain Research and Development* 32 (1): 51–61. https://doi.org/10.1659/MRD-JOURNAL-D-11-00091.1.

Prakash, Sanjay, ed. 1991. *Solar Architecture and Earth Construction in the Northwest Himalaya*. Sustainable Development Series 5. New Delhi: Har-Anand Publications in association with Vikas Pub. House.

Rizvi, Janet. 1998. *Ladakh: Crossroads of High Asia*. Delhi; New York: Oxford University Press.

Scott, James C. 2009. *The Art of Not Being Governed: An Anarchist History of Upland Southeast Asia*. Yale University Press.

Sudhalkar, Amruta Anand. 2010. "Adaptation to Water Scarcity in Glacier-Dependent Towns of the Indian Himalayas: Impacts, Adaptive Responses, Barriers, and Solutions." Thesis (MCP), Massachusetts Institute of Technology, Dept. of Urban Studies and Planning.

Tiwari, Sunandan, and Radhika Gupta. 2007. "Changing Currents: The Irrigation Practices of Leh Town." In *Recent Research on Ladakh*, edited by John Bray and Tsering Shakspo Nawang, 217–24. Leh, Ladakh: International Association for Ladakh Studies.

3   A changing climate

Ladakh was recognized for its extreme and relatively inhospitable conditions long before anthropogenic climate disturbances began to register in the area. And it is perhaps due to this extraordinary environment that the region also appears to be particularly vulnerable to the forces of climate change. Over the course of the past several decades, the Ladakhi experience has captured the public imagination, demonstrating not only the perils of communities on the front lines of climate change, but also the inventiveness associated with adaptive responses. Designers may find it useful to consider the experience of Ladakhis, as their response to climate change challenges marks a first wave of adaptive thinking, and the region stands out as a forerunner in the domain of climate change adaptation.

In addition to the climate and context, the finely tuned and extremely site-specific approach to subsistence farming in Ladakh suggests susceptibility to even small environmental changes (Heath 2015). The region's highly specific crops and farming practices lack the diversity that might make it possible to adjust to new norms. Climate change, with unpredictable shifts in weather that can impact the growing season, water availability and ambient temperatures, threatens to upend more rigid agricultural systems. Moreover, the gradual erosion of glaciers in the context of climate change also causes problems; without the meltwater provided by glaciers, an entire downstream farming system falls apart.

In addition to the pressures of climate change, Ladakh also suffers from environmental degradation associated with development. A growing population of Ladakhi people, an influx of immigrants and tourists, as well as the enormous presence of the army and the Border Roads' Organization all contribute to an increase in both resource use and pollution. This constellation of new human influences plays a part in the degradation of natural areas, the pollution of waterways, the use of limited resources such as water, food, wood and fuel, and the production of garbage (Dawa, Dana, and Namgyal 2000, 249). The two major environmental stressors of climate change and development thus give rise to conditions in Ladakh that demand mitigation, ideally, and at the very least, build a case for adaptation.

Climate change

Perhaps the most extreme manifestation of climate change in north India can be seen in the gradual recession of glaciers during the course of the last three decades. This erosion has reduced the supply of irrigation melt-water available to farmers and caused chronic drought for many subsistence-agricultural villages (Crook and Osmaston 1994; Grossman 2015). Moreover, this trend has only recently been linked to climate change, where elevated ambient temperatures and seasonal shifts in precipitation have resulted in a diminished snowpack and stock of glacial ice (Mingle 2015; Vince 2009). As journalist Jonathan Mingle notes, this warming appears to be maintained by a feedback loop caused by "declining precipitation, warmer winters, early-arriving spring, [and] reduced reflectivity on the glacier surface" (Mingle 2015, 221). Ladakh's changing climate represents a new era for the region, in which traditional weather patterns are no longer viewed as stable, and instability has become the new normal.

Because of the extreme climate and relative isolation of Ladakhi villages, surface meltwater is necessary for the production of crops, biofuels, and building material, and in turn, for the survival of humans and animals. Harvesting water from high glaciers and snowfields has in the past required extensive constructed canals; historically Ladakhi farmers "have made use of the barren alleviated semi desert conditions by cultivation through skillful irrigation; said to have been introduced in (the) tenth century" (Angchok and Singh 2006, 397). Now, as climate change impacts the quantity and seasonal availability of meltwater, farming households must adjust their traditional systems and practices.

Judicious agricultural practices have always governed resource use in the region. Historically, farmers in Leh District have sustained an intensively managed desert agriculture where limited water resources underpin an ethic of stewardship and also effectively curb growth. When author Janet Rizvi published her seminal book on Ladakh in 1983, she suggested that water scarcity would become the great obstacle to the region's development:

> Exercises in development, like irrigation schemes, tree planting, and exploration for ground-water, even if successful, will never scratch the surface of the problem. Water in abundance is simply not there and, unless the barrier of the Great Himalaya could be conjured away, never will be there. This landscape may be modified; it can never be transformed.
>
> (Rizvi 1998, 24)

Despite having experience living in an arid climate, the decoupling of precipitation events from seasonal norms has had a significant impact on agriculture in Leh District, where farmers have had to adapt to changing environmental conditions (Sangode et al. 2011).

As climate change impacts this already scant supply of water, farmers, engineers and other experts have scrambled to find new ways to extend resources in the region. Scholar Cai Heath documents three basic strategies for managing climate change using sustainable technologies that are already underway: passive solar techniques, artificial glaciation, and solar photovoltaic generation (Heath 2015). These projects may not all specifically connect to the topic of climate change, but they begin to chart a path forward under environmental pressures. A 2010 USAID report, for instance, suggests that the drainages in this region should be addressed by "well-planned management, conservation, and efficient use of the water people currently have available to them" (Malone 2010, 27). While one could argue that these suggestions have already guided water management practices in Ladakh for centuries, the development of new tools and techniques could help farmers manage climate insecurity.

Changing glaciers

According to the Leh District Statistical Handbook, due to the region's "arctic desert condition and (the) scanty rainfall of the district, irrigation depends on the eternal glaciers" (District Statistics & Evaluation Office 2013, 24). But many glaciologists, climate scientists, and journalists agree that the glaciers cannot be regarded as stable, much less eternal. New global weather patterns have caused large-scale changes to this landscape, resulting in a general decrease in glacial mass across the Himalayan range. Journalist Jonathan Mingle notes that the state of "Jammu and Kashmir has lost 20 percent of its total glacial mass in the past six decades" (Mingle 2015, 397) and cites a study of over 2,000 glaciers across the Indian Himalaya by India's Space Research Organization, which "determined that 75 percent are retreating, at an annual rate of 3.5 percent" (Mingle 2015, 222).

In their recent study on one glacier system in Ladakh, researchers Marcus Nüsser and Susanne Schmidt caution that "The high variability of glacier changes in this area does not allow for a simplified extrapolation of results on a regional scale" (Schmidt and Nüsser 2017, 119). Another study of the glaciers in the Zanskar region of Ladakh from 1962–2001 showed only minor decreases, compared to the rest of the Himalaya (Ghosh et al. 2014). Similarly, a USAID report finds the data inconclusive but acknowledges that "many of the glaciers in the Himalaya are indeed retreating, especially at the lower elevations in the eastern Himalaya" (Malone 2010, 2). The report recommends both mitigation and adaptation measures, to enable farming communities to continue to thrive even as water resources diminish. This document, which encompasses the larger scale of the trans-Himalayan range, also notes that the downstream effects of glacier retreat could be significant. As the system of glaciers that feed the Indus River diminish, communities that rely upon this river downstream will also be impacted. In this way, the water scarcity already felt in Ladakh could actually be amplified as one moves down in elevation (Malone 2010, 8).

While there has been in general a dearth of information with long time horizons of the glacial retreat in the area, trends occurring over previous decades are becoming incrementally visible. Indeed, while "data on the glaciers of the Himalaya, Karakorum, and Hindu Kush ranges are sparse and inconsistent" (Schmidt and Nüsser 2017, 107) glaciers are an important indicator of climate change, and on the whole are decreasing in the area. Despite varying predictions about the scale and pace of climate change in this region, it is clear that "glacier melt is occurring and will result, along with other hydrological changes, in impacts that will be felt in the basins of rivers that originate in High Asia" (Malone 2010, 25).

Changing glaciers can cause downstream problems such as "too much water (floods), too little water (droughts/increased aridity), or water at different times (more early in the growing season/less late in the growing season)" (Malone 2010, 25). This volatility is an acute problem for subsistence agricultural farmers, who rely on relatively stable weather and meltwater patterns to grow traditional crops, and to determine when to sow, water, and harvest their fields. Unlike farmers working with electricity, pumps, and other types of advanced agricultural technologies, Ladakhi villagers lack access to the tools that could otherwise provide a relative safety net. Instead they have relied almost exclusively upon the glaciers high above their villages to act as a storage device for water, and on regular seasonal weather patterns to release that water for irrigation use downstream.

A future under climate change

Because "Glaciers are remote and out of the farmers' control: their dependence on global climactic conditions determines the vulnerability of the farming system" (Pulselli and Pelliciardi 2012, 349). Adaptation measures in this context can include new techniques to conserve water, as well as shifts away from agriculture completely. In Leh District, a transition away from farming would have critical cultural and social ramifications. After all, this is a region in which "agriculture still remains the backbone of every village economy, engaging up to the 70% of the working force" (Pelliciardi 2013, 113).

In this dry desert environment, a number of villages have responded to a changing climate by reducing the historic footprint dedicated to agricultural production in order to size their fields to new limited supplies of water. Other farmers have resorted to selling off large numbers of their animals because adequate grazing land has diminished (Dame and Mankelow 2010). Farmers who opt to produce fewer crops or animals cut ties to their traditional livelihoods, relying instead on the government ration system, economic opportunities provided by tourism and newly built roads in the area, and employment beyond their ancestral village (Dame and Nüsser 2011; Pelliciardi 2013). In each of these cases, prolonged drought has altered the quality and character of village life, threatening the long-term viability of ancient cultural practices.

In an effort to improve food and economic security while retaining agricultural customs of the past, some villagers in this region have instead begun to experiment with new climate-adaptive design solutions. Many of the most visible physical alterations to these agricultural landscapes involve snow and ice, employing frozen design interventions that exploit existing freeze-thaw cycles to increase water catchment for farming. In essence, these frozen landscapes help to direct, trap, stockpile, and conserve scarce water resources. In combining culturally significant symbolism, traditional construction techniques, and new materials and design thinking, farmers in northern India are adopting an active role in water management and conservation. Other interventions include the construction of new architectural forms, the forging of relationships with non-Ladakhi allies, and protection measures for scarce resources.

For the relatively isolated region of Ladakh, climate-adaptive design solutions have primarily developed from the ground up. As early as 1998, scholar Janet Rizvi noted that development in this region is limited by a lack of funding and engineering support, where "Schemes to increase the amount of cultivable land through further irrigation can now only be taken up by those economic development agencies which have at their disposal huge quantities of money and technical expertise" (Rizvi 1998, 224). While many of the adaptive projects described in this book have been sponsored in part by non-governmental organizations such as the Leh Nutrition Project (LNP), Groupe Energies Renouvelables, Environnement et Solidarités (GERES), and Students' Educational and Cultural Movement of Ladakh (SECMOL), the vast majority of interest, participation, implementation, and oversight for these projects has occurred at the scale of the individual farm or village. Although significant funding, research, and workshop guidance has been contributed by these NGOs, the bulk of the effort to mitigate climate change-related drought through new design interventions has come from stakeholders at the community level.

The ground-up design thinking and action displayed in Ladakh runs counter to regional reports on climate change adaptation. For instance, in the USAID report of 2010, Malone notes that "Very few actual glacier-melt-related adaptation projects exist, and most of the projects tend to use large-scale, technological approaches" (Malone 2010, 3). This is simply not the case in Ladakh, where virtually all of the glacier meltwater adaptation projects currently in use rely on simple engineering practices that reroute water to collect in new physical formations (Nüsser et al. 2019).

Adaptive design responses

Ladakhi people have adopted numerous adaptive responses to make the most of limited resources in this high and dry environment. These design ideas bring together practices, tools and techniques that are some blend of new and old, low-tech and high-tech, and communal or individualized. They include design responses at the scale of architecture, landscape architecture and planning, as well as design ideas that harness opportunities within the region's social, political, economic, and environmental systems.

Architecturally, these solutions include the development of passive solar greenhouses and residential sunrooms, and the use of insulation, and large windows. Imported materials such as glass, tile, hardware, porcelain, mirrored glass, wall-to-wall carpeting, composite flooring, concrete, and CMU blocks have become increasingly common building products, even in Ladakhi villages not served by roads. Technologies such as those incorporating satellite dishes, Wi-Fi, electricity, indoor plumbing, pumps, and heaters have all been integrated into adaptive measures (Figure 3.1). Finally, historic building practices have been adjusted to meet new needs, as new development projects incorporate traditional masonry construction techniques, advantageous room arrangements, and principles honed through seasonal living patterns.

Adaptive landscape and planning initiatives include the reimagining of watersheds and agricultural landscapes, primarily as a means of improving access to scarce water reserves. New crops, and alternative mixes of crop types and animal husbandry reflect the changing constellation of economic opportunities and environmental constraints in the region today. Beyond these large-scale design interventions, many communities are working to retool social relationships, legal structures, and political systems to address climate change challenges. At every scale, Ladakhi design thinking marks this moment in history as one of creative transformation.

Living with change

Today climate change has become a major threat to the Ladakhi way of life. Unlike other forms of change – the ups and downs of state formation, the rising tide of globalization, or demographic shifts – this single force stands almost entirely outside of the ambit of Ladakhi agency. It is also a condition that connects Ladakh to the rest of the world, at times in unfair ways. The region's longstanding cultural knowledge and relative self-sufficiency may help it to adjust to these environmental pressures, but these two attributes are also susceptible to change under the pressures of climate change.

The knowledge and wisdom of Ladakhi stakeholders provides an important frame for adaptive development in the face of climate change. As communities develop new systems and ideas, flexible buildings and

Figure 3.1 A satellite dish sits in a yard in a village.

landscapes, and alternative modes of sharing and managing, they will necessarily draw upon their experience and expertise with the local climate and environment. Outside actors – whether NGOs, government agents, visitors or donors – would do well to also integrate this local viewpoint. After all, the placed-based knowledge that has been fine-tuned by Ladakhi people over the past ten centuries will likely offer valuable guidance into what could and should work in the future.

According to scholars Marcus Nüsser and Ravi Baghel, "there is a tendency among scholars and development practitioners to romanticize local knowledge systems and position them as antithetical to modern (and predominantly western) scientific knowledge." They note that local knowledge is necessarily born out of place, coming from the materials, resources, and environment of a particular site. While this site-specific knowledge provides the foundation for Ladakhi society, there has also been a long history of assimilation of ideas from other places, such as those brought forward by visitors, the state, and community groups. This has caused Ladakh to exhibit a "heterogeneity of local knowledge systems" (Nüsser and Baghel 2016, 192), which could be considered an asset in managing climate change adaptation in the future.

One connection between local knowledge systems and climate change can be seen in Ladakhi Buddhist communities (Figure 3.2). Here, the region's longstanding Buddhist values have become intertwined with environmental understanding, a connection that has the potential to amplify transformation under climate change. For instance, Scholar Sophie Day states that "Nature is controlled through ritual," where gods control glacial melt and thus the life in the valleys below (Day 1989, 57). According to scholar Andrea Butcher, in Ladakh under this belief system "The flow and condition of water is indicative of the success or failure of monastic ritual activity and human behavior; activities that have undergone significant transformations in recent years" (Butcher 2013, 110). In this line of thinking the effects of climate change might be attributed to behavior, and could also be improved through direct action. This conceptualization allows for intervention – and adaptation – to changing environmental conditions, through religious frameworks.

In adopting some agency for environmental management, Buddhist Ladakhis are completely prepared to participate in climate-adaptive design projects. But this empowerment can also clash with the perceived capacity of Ladakhi people at the state level. Indeed, Butcher also notes that "India's state models of development and economic planning, based upon modern governance paradigms, conflict with religious and ritual knowledge and practice" (Butcher 2013, 110). There is real friction between the agency assumed by local villagers for small projects and the role that state authorities absorb in the management of large infrastructural projects.

In the rapidly shifting environmental context of Ladakh, knowledge cannot be static, nor can it be entirely based upon traditional methods.

Figure 3.2 Stupas dot the landscape outside of villages in Ladakh, serving as a physical reminder of the connection between Buddhist religious values and the land.

Instead, the issues of resource scarcity, climate change, a growing and transitional demographic, the contested border and associated army presence, and the growth of tourism make the case for a similarly dynamic approach to adaptation. Ladakhi communities are adept at layering newer technologies, approaches, ideas and support onto more traditional forms of wisdom. This ability to transform practices using hybridity in the context of a shifting environment is perhaps the most useful adaptation approach currently in practice in Ladakh.

Together, the expressions of climate change evident in Ladakh today provide a layer of social, cultural economic, and environmental disruption above and beyond the impacts of the militarization of the area, challenges to border security, and the onslaught of globalization. The fact that villagers endeavor to develop solutions to this problem, and believe themselves to have agency in the mechanics of the environment, is in itself a reminder of the power and potential of Ladakhi cultural frameworks.

Finally, it is helpful to remember that Ladakh has been a site of transition over the centuries, adapting and flexing to a variety of environmental, political, economic, social and cultural shifts. Climate change is just one of many stressors in this landscape, and projections suggest that it will only increase in the coming decades. In this global landscape of climate change adaptation, Ladakh stands out as a forerunner for design guidance.

References

Angchok, Dorjey, and Premlata Singh. 2006. "Traditional Irrigation and Water Distribution System in Ladakh." *Indian Journal of Traditional Knowledge* 5 (3): 397–402.

Butcher, Andrea. 2013. "Keeping the Faith: Divine Protection and Flood Prevention in Modern Buddhist Ladakh." *Worldviews* 17 (2): 103–14. https://doi.org/10.1163/15685357-01702002.

Crook, John, and Henry Osmaston. 1994. *Himalayan Buddhist Villages: Environment, Resources, Society and Religious Life in Zangskar, Ladakh*. Bristol: University of Bristol.

Dame, J., and J.S. Mankelow. 2010. "Stongde revisited: Land Use Change in Central Zangskar." *Erdkunde* 64 (4): 69–73.

Dame, Juliane, and Marcus Nüsser. 2011. "Food Security in High Mountain Regions: Agricultural Production and the Impact of Food Subsidies in Ladakh, Northern India." *Food Security* 3 (2): 179–94. https://doi.org/10.1007/s12571-011-0127-2.

Dawa, S., D. Dana, and P. Namgyal. 2000. "Water Harvesting Technologies and Management System in a Micro-Watershed in Ladakh, India." In *Waters of Life-Perspectives of Water Harvesting in the Hindu Kush-Himalayas: Volume II*, edited by S.R. Chalise and M. Banskota, 235–59. ICIMOD. http://lib.icimod.org/record/22407.

Day, Sophie. 1989. *Embodying Spirits: Village Oracles and Possession Ritual in Ladakh, North India*. London: London School of Economics and Political Science.

District Statistics & Evaluation Office. 2013. *Blockwise Village Amenity Directory, 2012–13, Series 14*. Leh, Ladakh: District Statistics & Evaluation Office.

Ghosh, Swagata, A.C. Pandey, M.S. Nathawat, and I.M. Bahuguna. 2014. "Contrasting Signals of Glacier Changes in Zanskar Valley, Jammu & Kashmir, India Using Remote Sensing and GIS." *Journal of the Indian Society of Remote Sensing* 42 (4): 817–27. https://doi.org/10.1007/s12524-014-0368-6.

Grossman, Daniel. 2015. "As Himalayan Glaciers Melt, Two Towns Face the Fallout." Yale E360. March 24, 2015. https://e360.yale.edu/features/as_himalayan_glaciers_melt_two_towns_face_the_fallout.

Heath, Cai. 2015. "Climate-Friendly Development: Analysing Relationships between Community, Society and Government on Sustainable Technology Projects." *Ladakh Studies* 32 (January): 18–35.

Malone, Elizabeth. 2010. "Changing Glaciers and Hydrology in Asia." USAID. www.ehproject.org/PDF/ehkm/usaid-glacier_melt2010.pdf.

Mingle, Jonathan. 2015. *Fire and Ice: Soot, Solidarity, and Survival on the Roof of the World*. 1st edition. New York: St. Martin's Press.

Nüsser, Marcus, and Ravi Baghel. 2016. "Local Knowledge and Global Concerns: Artificial Glaciers as a Focus of Environmental Knowledge and Development Interventions." In *Ethnic and Cultural Dimensions of Knowledge*, edited by Peter Meusburger, Tim Freytag, and Laura Suarsana, 8: 191–209. Cham: Springer International Publishing. https://doi.org/10.1007/978-3-319-21900-4_9.

Nüsser, Marcus, Juliane Dame, Benjamin Kraus, Ravi Baghel, and Susanne Schmidt. 2019. "Socio-Hydrology of 'Artificial Glaciers' in Ladakh, India: Assessing Adaptive Strategies in a Changing Cryosphere." *Regional Environmental Change* 19 (5): 1327–37. https://doi.org/10.1007/s10113-018-1372-0.

Pelliciardi, Vladimiro. 2013. "From Self-Sufficiency to Dependence on Imported Food-Grain in Leh District (Ladakh, Indian Trans-Himalaya)." *European Journal of Sustainable Development* 2 (3): 109–22. https://doi.org/10.14207/ejsd.2013.v2n3p109.

Pulselli, Federico M., and Vladimiro Pelliciardi. 2012. "Emergy Evaluation of a Mountain Socio-Economic System and Traditional Agroproduction: A Case Study in Indian Trans-Himalaya." In *EMERGY SYNTHESIS 7: Theory and Applications of the Emergy Methodology Proceedings from the Seventh Biennial Emergy Conference*, edited by Mark Brown, 7: 347–56. Gainsville: Center for Environmental Policy.

Rizvi, Janet. 1998. *Ladakh: Crossroads of High Asia*. Delhi; New York: Oxford University Press.

Sangode, S.J., N.R. Phadtare, D.C. Meshram, S. Rawat, and N. Suresh. 2011. "A Record of Lake Outburst in the Indus Valley of Ladakh Himalaya, India." *Current Science (Bangalore)* 100: 1712–18.

Schmidt, Susanne, and Marcus Nüsser. 2017. "Changes of High Altitude Glaciers in the Trans-Himalaya of Ladakh over the Past Five Decades (1969–2016)." *Geosciences (2076-3263)* 7 (2): 27.

Vince, Gaia. 2009. "Glacier Man." *Science* 326 (5953): 659–61.

4 Development in Ladakh

The radical autonomy characterized by many of Ladakh's villages is not easily fabricated, and under current development trends this self-sufficiency is also gradually dissolving. In this ever-globalizing and hyper-networked age, major systems have become outsourced, from material goods to the production of information. Outsourcing also erodes notions of place, along with the knowledge and wisdom of a particular region. Even in the small villages of Ladakh, it is possible to see the ways in which individuals are now inextricably connected to each other, via relationships, products, and ideas, from across the globe.

While this interconnectedness is rapidly becoming a universal human standard and what scholar Neil Brenner calls part of the inevitable trajectory of 'planetary urbanization' (Brenner 2014), it also signifies a deeper system-wide vulnerability. Various types of outsourcing, for instance, can either fracture or disintegrate under the pressure of some sort of crisis. War, epidemics, or resource shortages caused by severe climate change all could swiftly precipitate a need for local services and products. The loss of local practices in advance of one of those catastrophic events signifies a loss of local autonomy, and the associated sustainable practices of that area.

Ladakh is a useful case study in this regard, for two reasons. The first reason is that the region has developed sustainable practices in a challenging physical environment, with minimal contact with the outside world, during the course of a thousand years. In this sense, the region harbors clear practices in self-sufficiency, resource husbandry, and long-term lasting power. It is a useful case study for a second reason, too: Ladakhi people stand on the front lines of climate change, experiencing the impacts of widespread drought and the erosion of many hundreds of glaciers. In this respect, Ladakh represents a region undergoing transition, and because it has been insulated from the rest of the world for so long, with very few external variables, it is possible to watch this change unfold. Unlike the theoretical climate futures projected in other parts of the world, Ladakh is already in the midst of major environmental change, with real experience to share.

Some combination of technique (in husbanding resources, using the land, and implementing sophisticated irrigation practices) and social organization (such as religious and family structure, consensus-based decision-making, and sharing) create a society in which self-sufficiency is possible. However, as variables in each of these realms shift, and as the surrounding context also changes (with the addition of roads, technology, education and the internet, for instance, or under natural forces such as climate change) self-sufficiency gets recalibrated. The case studies presented in this book demonstrate the techniques and social organizations used to leverage both long-standing sustainability practices and new adaptive measures. This design thinking preferences self-sufficiency in the local context, and for this reason it is useful to first explore the dimensions of power, development attitudes, and adaptation systems that characterize the region of Ladakh.

Ladakhi space and state power

In conjunction with its agricultural autonomy, the rugged mountain topography surrounding Leh District has effectively protected villages from political, social, and economic disturbance over the centuries. The Ladakhi people have historically lived at the state's periphery, exhibiting many of the same characteristics that scholar James C. Scott outlines in his study of Upland Southeast Asia and calls "adaptations designed to evade both state capture and state formation" (Scott 2009, 9). The remote and relatively inaccessible location of Ladakh appears to be central to the people's relative independence; if not intentional it has nevertheless been upheld as a central component of cultural and social identity in Ladakhi society, and self-sufficiency remains a cultural trademark and a point of pride. Scott suggests that such peripheral societies "may well have chosen their location, their subsistence practices, and their social structure to maintain their autonomy," and this framework helps to explain the independence exhibited by most Ladakhi village groups (Scott 1998, 8). In addition, Scott's articulation of the concept *'friction of travel'*[1] provides a framework for understanding how Ladakh's relative seclusion has effectively served as a buffer to both mainstream Indian and Western influences. While the difficult conditions of this high Himalayan region may challenge Western notions of comfort and livability, many Ladakhi people share an environment-centered skillset defined by extraordinary competence, and a lifestyle of cultivated political autonomy (Norberg-Hodge 1991).

It could be argued that residents of Ladakh did not experience the same level of post-colonial trauma that befell other parts of India, simply because the region itself is geographically remote, and thus buffered from the machinations of the state. During various periods of India's state-making, under colonial influence, and in the post-colonial era, Ladakh has maintained much of its own independent identity. The foundation for local government exists at the village level, which is born out in shared decision-making and traditional power agreements. Money and resources flow into the area via the political apparatus of the Union Territory of Ladakh, and, broadly, Indian laws are upheld. Yet, the power makeup and identity of Ladakhi villages appears to be more Ladakhi than Indian, and without much of the connective tissue for state building, such as effective taxation or national media control, the region can appear to function like an autonomous entity within the larger country of India.

As such, Ladakh could also be viewed as a new development arena, with a market to be claimed, a land to be territorialized, and people to be molded into subjects of the state (Figure 4.1). The Indian government maintains a strong national presence in this region today, primarily due to its overarching interest in border security. Major development projects in the area, such as hydropower projects, road construction, and the building of schools and airports, come from the Indian government. While other types of development projects have been initiated by NGOs and village groups, the largest infrastructural and physical building efforts have been provided by the state, and suggest overarching interests of national security and control. Perhaps the most telling indication of a plan to mold Ladakh into more of a state product can be seen in the 2019 repeal of Article 370 of the Indian Constitution, which subjected Ladakh to the authority of the Indian government, in an effort to shape the region's social and economic development.

In this context, top-down government-generated planning decisions tend to be out of touch with the largely invisible social, cultural and religious processes that guide most of the decision-making and daily activities throughout the region of Ladakh. Unlike the high authoritarian planning schemes that have characterized other Indian cities, the climate-adaptive design practices described in this book have instead been developed in and for Ladakh: they are knit into the Ladakhi cultural and social fabric, and are inextricably rooted to the region's environmental context. As a result, these design solutions have been finely tuned to the individualized needs of each community, and deployed on a case-by-case basis. Whereas larger, government-sponsored infrastructural projects might lack regional understanding of construction techniques, local materials, environmental conditions, or social frameworks, this set of climate-adaptive design solutions reflect the context and culture of Ladakh.

While social capital may not explicitly factor into the design of shared infrastructure projects, it can have major policy and urban planning implications. This consideration is perhaps even more relevant in tiny Ladakhi villages, where the meltwater from glaciers and snowfields has always been treated as a form of the commons, and shared infrastructure and buildings reflect the needs of the collective unit. In this region, subsistence agriculture practices demand a level of collaboration not often felt in more urban settings, as meltwater must be equitably divided amongst landholders, and farmers must work together to direct, store, and disperse this precious

Figure 4.1 The lush landscapes of old village land holdings clearly stand out next to new development projects beyond.

resource. Similarly, relationships that impact design opportunism, such as those forged with NGOs or visiting allies under the mantle of climate change adaptation, rely heavily upon social capital.

Resources and development trajectory

Resource scarcity has been a central factor in Ladakh's development over time. Limited available water, arable agricultural land, cooking and heating fuel, and wood for construction projects have constrained both overall population and development efforts. Likewise, these conditions have led to social and cultural norms that enable the efficient use of scarce resources. In recent years, a growing population and shrinking water reserves have contributed to the already limited pool of resources available in Ladakh. As the *Ladakh Autonomous Hill Development Council* articulated in their vision for 2025, this resource burden is only expected to grow (Ladakh Autonomous Hill Development Council 2005).

It is perhaps helpful to consider scholar David Harvey's research on resource use, in which he notes that constrained resources can actually lead to more widespread innovation: "By exhausting the social possibilities or depleting the natural resource base, a given mode of production will be forced to adapt and change in some way" (Harvey 2009, 201). In his writing Harvey also articulates nine basic human needs, including: food, housing, medical care, education, social and environmental service, consumer goods, recreational opportunities, neighborhood amenities, and transport facilities (Harvey 2009, 102). While this work references a European or North American context, the comparison with Ladakhi villages is telling. Most Ladakhi villages lack access to on-site modern medical care, educational opportunities and neighborhood amenities (such as libraries), a breadth of consumer goods, and extensive transportation options. The Indian government has worked to increase access to education and to improve access to transportation in the region, while also offering a ration to supplement village food production. To compare Ladakh to American or European development contexts highlights the gulf between opportunities and resources at these two respective poles. Nevertheless, a number of American and European scholars and visitors have identified Ladakh's development as a progressive model for emulation, as evidenced by the many books, articles and brochures written about Ladakhi exceptionalism (Norberg-Hodge 1991).

Narratives of misery

In Ladakh, tourism can be understood as a relatively new phenomenon. Although the region has been actively used as a crossroads of trade for centuries, the government effectively barred contemporary tourist access to the region from 1947 until 1974, and many areas still lack the amenities and infrastructure that facilitates economic opportunities and tourism in other parts of India. This limited access during the second half of the twentieth century is in part due to the region's physical location abutting the borders of Pakistan and Chinese-controlled Tibet, respectively, and the contested borders that those two countries share with India. It is also a remoteness undoubtedly shaped by the rugged terrain, high altitude, extreme weather, and lack of roads characterized by the trans-Himalayan mountain context.

While Ladakh is noted for its racking environmental conditions and limited year-round accessibility, it hardly fits into the "narratives of misery" (Demenge 2013) that visitors might superficially construe (Aggarwal 1997). Indeed, although isolated at times by climate and geography, today villages in Ladakh are by no means cut off from the rest of the world. Moreover, scholar Quentin Devers has conducted an archaeological study of ancient roads in Ladakh, in which he finds that prior to the seventeenth century, trade networks were likely even more diverse than in subsequent years, and also followed a variety of different mountain routes (Devers 2017). According to scholars John Crook and Henry Osmaston, movement has always been a defining feature of the region's economic framework, where "Caravan routes both within Ladakh itself and between Ladakh, Kashmir, Central Asia and Tibet have carried a vital if often slow moving trade" (Crook and Osmaston 1994, xxvii).

Today, the region is becoming more connected than ever before, with a growing network of roads, widespread cellular and television access, and in Leh town, direct daily flights to Delhi and other major cities. As the Ladakhi people explore formal and informal routes for contact with the increasingly globalized world, the region's physical buildings and infrastructure will also necessarily undergo change and transformation (Crowden 1997). Unfortunately, the 'narratives of misery' have more recently been used in this context as a tool to justify development interventions that are not only unnecessary, but also born out of misinformation (Demenge 2013, 58). According to scholar Jonathan Demenge, recent development in the area "illustrates how the notion of isolation is consciously manufactured and utilised, and a complex reality simplified and constructed into an object of knowledge – a poor, remote, backward, and isolated Lingshed[2] – to build the case for intervention" (Demenge 2013, 58). In this context, neo-liberal attitudes toward development and aid threaten to erode the traditions, daily lifestyle, and sophisticated social fabric that Ladakhi villagers have cultivated over centuries, all the while bolstering the state's hegemonic ideologies. As designers prepare to collaborate on projects in this region, they must first recognize the numerous and complex socio-spatial influences on development in the area, including the formal and informal systems of governance, the physical context, and the power relationships that have heretofore shaped Ladakhi society.

Scholars and visitors have helped to both disseminate and dissolve these notions of rural deprivation. According to scholar Ravina Aggarwal, "Ladakh has been represented in dual terms, as a surviving remnant of the glories and mystic secrets of an unsalvageable Tibet and as a primitive wilderness at the fringes of the Indian sub-continent" (Aggarwal 1997, 23). These stories obstruct a reading of an authentic contemporary place, with attendant complexity and diversity. Aggarwal also notes that this kind of representation hampers authentic engagement, where locals become "quarantined into the exotic confines of prison-like spaces and prison-like modes of thought which academic territorialism and imagination have defined and concretized" (Aggarwal 1997, 21–2). Similarly, Ladakh's designation as a "Little Tibet" reduces and totally undervalues the unique cultural positioning of the Ladakhi people, most of whom would never consider themselves Tibetan (van Beek 2003). This Tibetan idealism is reinforced by scholarly disdain for written Ladakhi, which instead preferences Tibetan script and formal language.

Guidebooks and tourism brochures often highlight the extreme remoteness of the region in an effort to sell a vision of a landscape that also harbors a well preserved cultural product. As Ravina Aggarwal notes, "Most visitors to Ladakh carry with them this romantic notion of an idyllic land, eclipsed from time and space" (Aggarwal 1997, 22), and this imagery does, indeed, stem from the messaging used by the local tourism industry. She argues instead for a rehabilitative anthropological view of Ladakh, where people and culture are fluid, mixed, and diverse, never fixed in time or space. Similarly, scholar Martijn van Beek makes a case for a new conception of Ladakhi modernity that is a local project, driven by the needs and desires of local stakeholders (van Beek 2003). Indeed, somewhere between romantic notions of a lost Shangri-La, and the development discourse's narratives of misery, Ladakh instead offers a complex story about people and place in transition.

Adaptation in Ladakh

Ladakhi villages are changing to accommodate new social and cultural norms, as well as in response to the growing environmental pressures caused by climate change. This transformation necessarily occurs within the context of a relatively autonomous and self-sufficient cultural fabric. According to scholar Sally Daultrey,

> In Ladakh, living with change means doing more with less – not because there is suddenly less, but because in a remote, resource-constrained, ecologically sensitive area, working with what is available is the first and sometimes the only option. Unlike computer parts, tourists and food supplies, adaptation does not arrive on a plane from Delhi.
>
> (Daultrey 2010)

In all its complexity, the relative inaccessibility and self-sufficiency of present-day Ladakh provides a model for independent, sustainable community organization. Many of the products used in remote villages

Figure 4.2 A homemade wheelbarrow in Ladakh.

Figure 4.3 Villagers collect dung for fuel in the pastoral landscapes of Ladakh.

have come from the surrounding region, or are made on site (Figure 4.2). Residents in these villages tend to know how to manage for themselves and their households in this environment – how to provide food and fuel, for instance, within the carrying capacity of the land (Figure 4.3). And at the village level, communitarian values have forged strong social bonds and facilitated the long-term management of the land.

As Ladakhi villagers adapt to a changing climate, and an increasingly globalized world, they have intentionally blended old and new ideas. According to Daultrey, "By selecting the best available technology and information, (and) fusing it together with indigenous capabilities and knowledge, Ladakh is demonstrating the true meaning of adaptation in a rapidly changing world" (Daultrey 2010). It is possible that adaptation ability, and the inventive hybridization of traditional and current systems, could steer the course of Ladakhi development in future decades.

In their study on Ladakh, Sally Daltrey and Reuben Gergan identify the characteristics of adaptation in the region, where the specific attributes of society shape transformative potential. According to these scholars,

> Adaptation in Ladakh may be defined as: using existing social structures that recognize cultural diversity to achieve cooperation in the pursuit of development objectives; applying creativity and knowledge to use resources judiciously; overcoming short-termism; integration and adaptation of technology to suit the environment, choosing the right technology at the right scale: choosing technologies which match system dynamics (decentralized technologies work well in decentralized communities); harmonizing state-level policies with micro-development plans and the needs of highly dispersed communities; interaction with global knowledge through traditional routes; and utilizing experience and knowledge of a new generation of Ladakhi citizens.
>
> (Daultrey and Gergan 2011, 10)

In Ladakh, the pressure to develop along with the Western world, to interface with the rest of the world, and to participate in the global economy is perhaps a more visible driver of adaptation even than climate change. Planetary urbanization has already impacted trade, building standards, and the types of employment sought by younger generations. New forms of development, in any case, is at once desired by many Ladakhi residents and pressed for by the state, unleashing a host of climate-adaptive design responses as well as the more typical indicators of progress: the building of roads, technology adoption, and ever-increasing access to global cultural products.

Notes

1 James C. Scott describes the *friction of travel* as a method for considering the added dimension of difficult terrain in access calculations (Scott 2009).
2 Lingshed is the name of the Ladakhi village referenced by Demenge.

References

Aggarwal, Ravina. 1997. "From Utopia to Heterotopia: Towards an Anthropology of Ladakh." In *Recent Research on Ladakh 6: Proceedings of the 6th International Colloquium on Ladakh, Leh, 1993*, edited by Henry Osmaston and Tsering Nawang, First Asian Edition, 21–8. Delhi: Motilal Banarsidass Publishers.

Beek, Martijn van. 2003. "Imaginaries of Ladakhi Modernity." In *Proceedings of the Tenth Seminar of the IATS*, 11: 163–88. Brill.

Brenner, Neil J., ed. 2014. *Implosions/Explosions: Towards a Study of Planetary Urbanization*. Berlin: Jovis.

Crook, John, and Henry Osmaston. 1994. *Himalayan Buddhist Villages: Environment, Resources, Society and Religious Life in Zangskar, Ladakh*. Bristol: University of Bristol.

Crowden, James. 1997. "Development and Change in Zangskar, 1977–1989." In *Recent Research on Ladakh 4 & 5, Proceedings of the Fourth and Fifth International Colloquium on Ladakh*, edited by Henry Osmaston and P. Denwood, First Asian Edition, 271–80. Delhi: Motilal Banarsidass Publishers.

Daultrey, Sally. 2010. "Adaptation on the Roof of the World." *The Center for Design and Geopolitics* (blog). December 30, 2010. http://designgeopolitics.org/blog/2010/12/adapatation-on-the-roof-of-the-world/.

Daultrey, Sally, and Reuben Gergan. 2011. "Living With Change: Adaptation and Innovation in Ladakh." *Climate Adaptation*. Available: www.yumpu.com/en/document/view/25089047/living-with-change-adaptation-and-innovation-in-our-planet.

Demenge, Jonathan. 2013. "The Road to Lingshed: Manufactured Isolation and Experienced Mobility in Ladakh." *HIMALAYA, the Journal of the Association for Nepal and Himalayan Studies* 32 (1): 51–60.

Devers, Quentin. 2017. "Charting Ancient Routes in Ladakh: An Archaeological Documentation." In *Interaction in the Himalayas and Central Asia: Processes of Transfer, Translation and Transformation in Art, Archaeology, Religion and Polity*, edited by Eva Allinger, Frantz Grenet, Christian Jahoda, Maria-Katharina Lang, and Anne Vergati. Vienna: Austrian Academy of Sciences Press. www.academia.edu/39101653/Charting_Ancient_Routes_in_Ladakh_An_Archaeological_Documentation.

Harvey, David. 2009. *Social Justice and the City*. Revised edition. Athens: University of Georgia Press.

Ladakh Autonomous Hill Development Council. 2005. *Ladakh 2025: A Road Map for Progress and Prosperity, Ladakh 2025 Vision Document*. Leh, Dadakh: Ladakh Autonomous Hill Development Council. https://cdn.s3waas.gov.in/s3291597a100aadd814d197af4f4bab3a7/uploads/2018/06/2018061732.pdf.

Mann, Rann Singh. 1986. *The Ladakhi: A Study in Ethnography and Change*. Delhi: Anthropological Survey of India.

Norberg-Hodge, Helena. 1991. *Ancient Futures: Learning from Ladakh*. Sierra Club Books.

Scott, James C. 1998. *Seeing Like a State: How Certain Schemes to Improve the Human Condition Have Failed*. Yale University Press.

Scott, James C. 2009. *The Art of Not Being Governed: An Anarchist History of Upland Southeast Asia*. Yale University Press.

5   Challenging design engagement

While much of the climate-adaptive design transformation in Ladakh has come from within, outside design assistance has also acted as a vital source of support for new projects. In these cases, visiting designers pair up with either villages or specific stakeholders to launch built projects. As a climate-adaptive design strategy, international collaborations have the potential to quickly affect change in the region, leveraging the resources, knowledge and energies of visiting designers. This technique is also potentially fraught, as it places Ladakhi people in a position to receive design aid that may or may not wholly be welcomed. Outside design assistance may continue to be incorporated in the development of future climate-adaptive projects, and indeed it stands out as a possible climate-adaptive design strategy in and of itself. However, this work would benefit from a critical consideration of power roles in the context of collaborative practice.

Background

When the Indian government officially opened Ladakh to foreign visitors in 1974, dozens of mountaineers, geologists, Buddhist scholars, and tourists ventured into the region. This number grew steadily over the years, and with it, so too did the relationships that would foster ongoing humanitarian sponsorship and aid. Today many of the villages have benefitted from design and planning assistance donated by international sponsors, NGOs, and visiting volunteers.

Virtually all of the projects that have been built in this context reflect a desire to improve the living conditions of Ladakhi villagers, to improve environmental stewardship, and to share new technologies and ideas in a climate of cross-cultural exchange. However, while visiting designers may bring valuable experience to Ladakhi development projects, their deep-seated professional norms, inevitable cultural distance, and unrealistic construction expectations threaten successful outcomes. This chapter identifies this source of support as a technique for bolstering climate change adaptation efforts, explores the friction of cross-cultural aid work in Ladakh, and suggests a way forward for practitioners of public-interest design engagement.

The problem with international pro-bono practice

In an effort to connect design services to Himalayan communities in need, foreign architects, urban designers, and landscape architects have, over the course of the last several decades, participated in a wide-ranging array of pro-bono design projects. These community-oriented efforts can benefit from the guidance of an external design team, whose valuable design insight and vision, expertise with building technologies and innovative materials, and links to both funding sources and international media outlets would otherwise prove inaccessible. In addition, these public service projects often have the effect of catalyzing new development around a highly visible built project, demonstrating the potential for cascading benefits caused by a single design intervention and calling attention to the climate challenges of the region.

However, the "gift" of design service also may carry significant drawbacks, well understood in post-colonial and post-disaster contexts. Design teams bring their own interests and agendas to a pro-bono building project, all too often shrouded in a mantle of aesthetic superiority and design expertise. This type of engagement conjures up some of the same attitudes that drove the high modernist schemes described by the American political scientist and anthropologist James C. Scott, which "despite their quite genuine egalitarian and often socialist impulses," tend to show little confidence "in the skills, intelligence, and experience of ordinary people" (Scott 1998, 341). Such hierarchical distancing threatens successful project outcomes; without stakeholder investment and a climate of mutual respect, designers too easily overlook important functional, cultural, and aesthetic cues.

Physical distance also jeopardizes collaborative schemes. Without firsthand knowledge of, and lived experience near a project site, many visiting designers fail to comprehend the unique climactic, social, and cultural context that would necessarily ground a successful design intervention (Perkes 2009). This physical remove becomes even more evident once the project is underway. Without a lasting on-site presence, and a real stake in the project, absentee designers may struggle to bring a project to fruition. Visiting design teams may begin projects abroad and then depart before they are finished, or if the project is completed, could be conspicuously absent when components need maintenance or servicing. Even the most experienced designers practicing in international contexts recognize the limitations of working abroad and the difficulty of shepherding a successful process without a consistent presence on the ground.

It could be argued that designers who work on pro-bono projects in foreign countries have an ethical responsibility to model the very best practices for international engagement. Because these efforts tend to be highly visible, community-oriented, and outside of the normal scope of conventional work at that location, public-interest design projects necessarily have more at stake. Moreover, because many of these interventions intentionally target vulnerable or marginalized communities, the power dynamic between designer and client can be lopsided, and this imbalance must be taken into consideration in order to ensure equitable transactions. This is not to say that service work should be abandoned; rather, as public-interest design practice becomes a more common form of volunteerism in Ladakh, professionals will need to grapple with both the positive and negative implications of their work. Skillful practitioners who move forward with full awareness of these challenges are needed more than ever, if only to set the tone for appropriate practice.

Towards a selfless practice

From the swagger of Ayn Rand's character Howard Roark in *The Fountainhead* to the most prominent Starchitect heroes, hubris has become one of the defining characteristics of the architectural profession in Western popular culture (Deamer 2014). This conceit is reinforced by an architectural academy that celebrates the production of radical, unmoored creations (Fisher 2012) and increasingly complex modes of modeling, representation, and distribution. While courageous creative expression is central to the teaching and practice of design, more grounded modes of engagement must also be recognized and cultivated. One model comes from educator Sergio Palleroni, who integrates public-interest design into the classroom at Portland State University and has created a process for helping students to dismantle their architectural egos. In preparation for an international pro-bono design/build project, students first spend an entire "year at their home institution, investigating the physical and cultural characteristics of the client community, documenting the site and programmatic requirements, and engaging in group charrettes" (Palleroni 2011, 227). This preparation helps the students to fundamentally connect their conception of design practice to clients and context, so that their work on site is more appropriately tethered to the people and place that they intend to serve.

India's public interest architecture

The concept of selfless service is not a new idea for the discipline of architecture, nor is it exceptional in India. In his book on the life and work of architect Laurie Baker, the writer Gautam Bhatia suggests that an architect's contribution to society could "be looked upon as the public's perception of him as a socially responsible professional, and his work as a socially responsible act" (Bhatia 1994, 66) and he argues that Baker embodied these characteristics. Laurie Baker was just one of the foreign-born architects who helped to shape India's public-interest design repertoire, and his legacy offers an inspiring model for future practitioners. Moreover, just as the desire to contribute to an ethical practice is a goal shared by many practitioners

(Deamer 2014), in India it also suggests one of the time-tested routes to achieving appropriate development.

India provides a unique set of challenges for public-interest design engagement, especially by foreigners. In India, architectural aid work is necessarily both a product of, and a response to, the particularly challenging context of the country's struggle to reinvent itself after colonial rule. In this post-colonial nation, architecture has served as a both a symbol of oppression and emancipation, and the physical attributes of space can often express deeper meaning.

Beyond these material remnants, the practice of design advocacy by outsiders has itself a checkered past. As James C. Scott notes, the precedent for architectural aggression has already become well established in India, where the projects of high-modernist statecraft were often led by teams of foreign designers. Perhaps the most controversial foreign architect working under the pretext of social betterment in India was Le Corbusier, who, in conjunction with Albert Mayer, Matthew Nowicki, Pierre Jeanneret, Jane Drew, Maxwell Fry, M.N. Sharma, A.R. Prabhawalkar, B.P. Mathur, Piloo Moody, U.E. Chowdhury, N.S. Lambda, Jeet Lai Malhotra, J.S. Detghe, and Aditya Prakasha, developed the masterplan for the new city of Chandigarh. The plan was an example of foreign-sponsored high modernism; a radical departure from both traditional and colonial aesthetics, as well as functional and climactic responses. In the design of this modern city, Scott argued that "the progenitors of such plans regarded themselves as far smarter and farseeing than they really were and, at the same time, regarded their subjects as far more stupid and incompetent than they really were" (Scott 1998, 393). While such neocolonial attitudes are arguably less pronounced in India today, the entrenched scars from this period cannot be overlooked or brushed away.

At that time, dominant development attitudes associated notions of expertise and value with design credentials. Moreover, visiting designers such as Le Corbusier tended to defend the primacy of their role within the context of providing services. According to Scott, "Believing that his revolutionary urban planning expressed universal scientific truths, Le Corbusier naturally assumed that the public, once they understood this logic, would embrace his plan" (Scott 1998, 114). Le Corbusier's approach to design service in India exemplifies many of the professional attitudes of his time: the all-knowing expert architect could override local preferences and conventions in an effort to import so-called best practices from another place. Such neo-colonial thinking continues today in other parts of India; foreign-sponsored public-interest design projects have been packaged through a well-intentioned but imported design visions, in the process reinforcing a convenient narrative of architect-as-expert.

Architect Laurie Baker offers a powerful counter example to this type of foreign design influence in India. Although he was born in the United Kingdom, Baker lived and worked in India for his entire adult life, eventually becoming an Indian citizen. Baker made public-interest architecture his life's work, developing a vernacular-oriented practice that served and represented all types of people, integrated regional methods, and prioritized low-cost sustainable design suited to local climate and context. According to his biographer Gautam Bhatia, "Baker's work has often been referred to as the architecture of marginality as his designs make optimum use of available funds and materials" (Bhatia 1994, 48), and indeed the long list of projects he produced in India support this contextual thinking. In drawing inspiration from the local vernacular, and his attendant commitment to building projects that supported the poor, Baker set the tone for a more enlightened altruistic architecture practice in India.

Design engagement in Ladakh

North India has a relatively short record of international design engagement as compared to rest of India, primarily due to its remote location, which only recently opened to foreigners. The extreme inaccessibility of Ladakhi villages restricts the technologies, building materials, and tools that can be incorporated in construction projects. In addition, many villages lack the basic provisions and infrastructures that would facilitate Western-style development, such as reliable electricity for powering tools, or viable roads for material deliveries. Regardless of these difficulties, visiting foreigners have supported small construction projects since the 1970s; a constellation of altruistic outcroppings incorporating new materials, technologies, and forms.

For design practitioners working in the remote context of Ladakh, the realm of public-interest architecture occupies challenging theoretical, practical and ethical terrain. Because this is a remote and rugged landscape, far from the cultural and physical realities of their home experience, it is a practice that merits both study and restraint. Designers need to adapt to new material sourcing, technical skills, construction standards, and cultural norms. As outsiders, visiting designers must embed themselves in the unique context of the region, preferably under the direction of local stakeholders.

Working both with organizations and independently, a host of well-meaning foreign design teams have participated in aid work in the region. These projects range from tree-planting to the development of clinics and schools. Professional firms, such as the international engineering group *ARUP*, have sponsored substantial long-term building efforts in the region (Brislin 2008). Non-profit groups such as *Architectes Sans Frontiéres* and *Groupe Energies Renouvelables, Environnement et Solidarités* have worked on building projects at smaller scales throughout Ladakh, building schools and clinics but also spreading new technologies and supporting workshops (Figures 5.1 and 5.2). Student groups have visited the region to travel and volunteer for decades, such as *The Institute for Village Studies, Where There Be Dragons, BaSiC Initiative*, and *Studio Ladakh*. Many of

Figure 5.1 The exterior of the Lamdon Model School, built and designed by the
          group Architectes Sans Frontières.

these visiting designers have cultivated strong and lasting partnerships with
community members, producing mutually-beneficial work over time.

Both professional firms and service organizations facilitate collaborative
climate-adaptive work in Ladakh. Architecture firms may take on pro-bono
projects as a form of disciplinary service; as a reflection of the interests and
aims of a single practitioner; or as a means of cultivating relationships in
new contexts. Many independent groups have also emerged with the sole
purpose of addressing humanitarian design engagement; these collectives
typically are organized as non-profit groups or NGOs. Like the now-defunct
organization *Architecture for Humanity*, these groups serve as matchmakers
and managers; facilitating challenging project management in international
contexts, improving visibility through targeted public relations, cultivating
donors and simplifying the giving process by retaining a non-profit status,
and sustaining multi-year service campaigns even as internal personnel
changes. Groups such as *Architectes Sans Frontiéres* have moved to the forefront of this effort in Ladakh and are strategically working on new environmentally sensitive projects. Current practitioners now treat architectural aid
as a specific genre of skilled work; yet whether pro-bono or fee-based, these
firms and organizations still operate in a largely unregulated professional
sphere (Charlesworth 2014; Wickersham 2014).

Figure 5.2 The interior of the Lamdon Model School in Zanskar, designed and built by Architectes Sans Frontières.

Moving forward

As designers continue to develop best practices for public-interest design work, difficult questions will need to be addressed. How can outside designers (with their so-called expertise, energy and resources) effectively integrate with the client designers (who have a much deeper awareness of their core issues and will ultimately need to live with the results)? Is there an inherent power imbalance that should be recognized in cross-cultural work? Whose values and beliefs will be reflected in the built project? Is the work equitable? One would hope that with the appropriate project framing, communication, and expectation management, foreign practitioners and local stakeholders could come to complement each other in the co-production of new space. Not only do project outcomes tend to improve through authentic collaborative efforts, but in the context of international aid work, anything less cooperative could be unethical.

One of the central challenges to successful public-interest design projects in Ladakh is in managing exposure to the climate, context and culture of the host client. Many visiting designers can only offer a limited amount of time, on site, in their host country. In north India, the relative isolation of small villages, and difficulties in gaining access to people, resources and a site effectively compound this scheduling challenge (Figure 5.3).

Many people who want to help in a foreign context are inevitably out of touch with the realities of the field conditions. Taking time to spend on site, working directly with a community, is one way to gain familiarity with a place, a people and their needs. According to Sergio Palleroni, time on site can help to improve the built product made by students: "After the students become familiar with local building technologies and resources, they are better able to finalize the details in ways that are pertinent and resonate with local culture" (Palleroni 2011, 229). Moreover, the time spent with community members helps to build trust and investment with all parties.

Throughout engaged design efforts, designers should work with the community as a partner. This creates an environment in which co-design is used to produce work. Scholar Gabriel Arboleda advocates for moving beyond co-design to place the visiting designer at the bottom of the decision-making hierarchy in order to empower clients to drive the design process (Arboleda 2020). Regardless, there is an ethical imperative for designers to not just open the door to let clients view the design process, but to ensure that the process itself is open, inclusive, and equitably engaged. Visiting designers should have a hard look at their motivations and practice with regard to international pro-bono practice. Collaborators may bring their own stereotypes and cultural frames, seeing the populations they work with as incapable of participating in expansive ways.

While the Ladakhi people have been working alongside foreigners to envision development projects since the region officially opened to outsiders in the mid-1970s, the local culture of hospitality might ultimately cause a concession to an outsider's viewpoint. According to journalist Jonathan Mingle, here "given their agrarian-communitarian history, people tend to value avoiding conflict over social innovation" (Mingle 2015, 274). In this sense, the Ladakhi people might have a strong sense of what they want or

Figure 5.3 A stakeholder reviews design products for the Tungri Nunnery, brought by a visiting French architect.

need, but will be willing to shift positions or even accept something totally outside their interests for the sake of broader social harmony. This climate of assent creates a major hurdle for even the most selfless design practitioners; unless they can truly participate in a process of co-design, the needs of clients might not be heard.

There is another argument against public-interest design engagement in international contexts: if public buildings are sponsored by foreigners then that work effectively releases the state from any duty to protect and provide for its subjects. This is a real concern in Ladakh, where despite the radical autonomy of most people, villages routinely look to international NGOs to develop much needed public infrastructure, and buildings such as schools or clinics. This aid can act to extend the number of opportunities for Ladakhi people, as in the case of the sheer number of students attending high school sponsored by outside benefactors. However, without this type of intervention, the government could potentially be held more accountable, to provide necessary services, support and opportunities to Ladakhi state subjects.

Thinking ahead

In the context of Ladakh, altruistic design work is becoming an increasingly popular endeavor; this is a product of heightened professional interest in humanitarian design, the growing numbers of foreign visitors each year and an attendant rise in volunteerism by this group. While this trend illuminates the enthusiasm for humanitarian work in the design disciplines, it also calls into question the appropriateness of volunteer aid in the development of critical infrastructures, artifices, cultural amenities, and other projects typically considered to reside within the domain of the state. Moreover, because this work occurs in the public realm, it also prompts a broader discussion about the moral and ethical implications of the work, a call for a set of guidelines for the physical standards of service projects, and an appeal for accepted, contextually appropriate modes of design engagement.

Architectural activism, after all, can unwittingly reinforce the hegemonic ideals of the state. As James C. Scott suggests, this practice already has roots in post-colonial India, where the heavy hand of high modernist projects once sought to apply the rational architecture values outlined by the CIAM (Scott 1998). While designers engaging in altruistic aid projects must have good intentions, it is possible that their work ultimately embodies the colonialist thinking of the past.

Critical reflection upon foreign design engagement provides an opportunity to re-evaluate the practice, interests and motivations of public-interest architecture, but not an excuse to write it off. Gautam Bhatia's provocative suggestion that "Unless architecture transcends its traditional scope, (i.e. working for the rich or for large institutional projects) architects will do incalculable damage to the environment and to the existing patterns of society" serves as a reminder that the practice of architecture also includes moral and ethical responsibilities (Bhatia 1994, 24). Other architectural theorists argue that the existing system for a market-based architectural practice is inherently flawed, as it excludes the majority of the world's population and ignores the issues of poverty, disaster, and humanitarian aid (Charlesworth 2014). Indeed, Patrick Coulombel notes that "it is appalling that architects remain uninterested in and out of touch with building for the most vulnerable and impoverished people" (Coulombel 2011, 286), and Jason Pearson suggests that

> Design professionals – from graphic designers to planners, architects to product designers – are in a unique and strategic position to influence the ongoing creation of the images, objects, and environments with which we surround ourselves. With that unique position comes a responsibility to understand and respond to a larger public good.
>
> (Pearson 2002, 5)

If designers have a professional obligation to engage in service, then the question is not if, but how, they might compose this altruistic work. This chapter highlights an emergent idea in architectural aid circles today: it is the suggestion that designers must conceptually and practically move beyond the practice of "gift-giving" to focus instead on the co-production of service projects. This view holds that lasting, authentic partnerships between expert designers and expert stakeholders will set the tone for more equitable international advocacy work.

Ladakh's geographical isolation from the rest of the developing world has, until recently, preserved many of the cultural, social, and physical characteristics of traditional village life. However, as the number of foreign visitors increases, as subsistence agricultural practices wane, as new roads are completed, and as access to the internet and television grows in these remote regions, Ladakh's distinctive cultural, social, economic and environmental characteristics will increasingly bend toward more globalized uniformity.

As the inevitable tide of globalization sweeps through Ladakh, designers – especially those coming from practices abroad – have an opportunity if not an ethical imperative to consider the unique climate and context of their sites of engagement. In so doing, visiting designers might draw from the values outlined in Kevin Frampton's *Prospects for a Critical Regionalism* by embracing social and cultural frameworks as well (Frampton 1983). Future public-interest projects must reflect an awareness of power relationships and social fabric, environmental conditions and resource scarcity, building wisdom and material use, as well as the aspirations and needs of village people. In producing work in Ladakh, foreign design teams must recognize these vital shaping forces while simultaneously embracing new forms and ideas, technologies, safety standards, and programmatic elements. Contextually appropriate development, after all, recognizes the importance of finding the right fit, rather than expressing an allegiance to the relics of the past.

If, as David Harvey and Susanne Langer suggest, architecture is "an ethnic domain" (Harvey 2009, 31) then international public-interest design practice also should reflect local social constructs. Such involvement suggests a hybrid, neo-vernacular approach for pro-bono design practice: one in which time-tested traditions, building wisdom, materials and environmental sensibilities are paired with new technologies, higher standards for safety and health, and the aspirational thinking of local stakeholders. By referencing deep cultural links, visiting designers have an opportunity to illuminate a community's independence, rather than reliance, on external sources.

This connection between cultural expression and sovereignty is even more pronounced in international practice, where the forces of neoliberal aid practices tend to dominate development work. Many of the neoliberal ideals that have heretofore characterized architectural aid abroad – designers swooping in to save a community that cannot save itself – appear to be outmoded as well as patronizing. Accordingly, public-interest design practitioners must be careful to avoid, in Jay Wickersham's words, "acting as unwitting tools of inequality and repression" (Wickersham 2014, 33). In recasting international practice, skillful public-service design projects can provide a useful vehicle for what Peter Evans describes as "counter-hegemonic globalization" (Evans 2008). Public-interest design engagement offers an opportunity to transcend the traditional roles in design and elevate architecture to a form of public service.

Naturally, public-interest design practitioners will only realize such lofty desires when their work is conceived in partnership with local stakeholders. This chapter suggests that designers can draw from the rich environmental and cultural context of north India to create useful, appropriate, and forward-thinking design interventions in Ladakh. Moreover, this design engagement is valued by community members not so much because it represents noble charity or access to state-of-the-art space, but instead because the work reflects the authentic and lasting partnerships that first ushered it into existence.

In crafting a roadmap for future public-interest practice in north India, designers would do well to reference Ladakh's core operating values: useful design engagement stands on the foundation of genuine altruism, communitarian accord, and environmental harmony. One would hope that with the proper setup, framing and communication, aid groups could come to complement local stakeholders. Not only do the outcomes tend to be better through real collaborative efforts, but the alternative is unethical. In the process, the discipline of architectural could create new value by responding to the mounting environmental, social, and political challenges of our rapidly urbanizing planet.

The belief that design service should benefit all people, rather than only those in the top socio-economic tiers, is hardly a radical idea. This message has become the rallying cry trumpeted by many of the altruistic firms and service-based organizations operating in the sphere of public-interest design. On the surface, these groups and their social impulses appear to be unequivocally beneficial. Indeed, many of the characteristics of public-interest design projects embody a full slate of co-benefits, for the client, the designer, the environment, and more. But a more critical examination of public-interest architecture reveals a host of under-represented issues, ranging from power imbalances to aesthetic decision-making, and levels of engagement to underlying motivations. These unsettling gaps in knowledge and action, particularly in the context of international aid work, have prompted a host of questions about this practice (Nussbaum 2010).[1]

As many scholars have already noted, understanding the "hazards of charity" (Rieff 2003) is a critical and necessary exercise for aspiring practitioners of public-interest design. Now, more than ever, the emergent domain of "do-good" design would benefit from a more nuanced understanding of both the advantages and disadvantages associated with this practice, and a clear set of guidelines for appropriate engagement. Educator Jay Wickersham argues for a new code of context, noting that "As architects debate these questions about design, sustainability, and ethics in international projects, we may start to develop a shared set of principles and behaviors, which can help guide global practice in the future" (Wickersham 2014, 35). This book illuminates some of these guidelines within the unique working context of north India, and addresses several examples of public-interest design engagement in the climate-adaptive design projects of Ladakh.

Note

1 Educator Jay Wickersham, for instance, has asked: "Are architects helping to strengthen and develop the economies of host communities, or are they acting as unwitting tools of inequality and repression?" (Wickersham 2014, 33).

References

Arboleda, Gabriel. 2020. "Beyond Participation: Rethinking Social Design." *Journal of Architectural Education* 74 (1): 15–25.

Brislin, Paul. 2008. *Unified Design*. Chichester, England: Wiley & Sons, Incorporated, John www.biblio.com/book/unified-design-arup-associates-staff/d/1270925723.

Charlesworth, Esther. 2014. *Humanitarian Architecture: 15 Stories of Architects Working after Disaster*. 1st edition. New York: Routledge.

Coulombel, Patrick. 2011. "Open Letter to Architects, Engineers, and Urbanists." In *Beyond Shelter: Architecture and Human Dignity*, edited by Marie Aquilino, 291. New York: Metropolis Books.

Deamer, Peggy. 2014. "Invitation to a Dialogue: Less Ego in Architects." *The New York Times*, August 2, 2014. www.nytimes.com/2014/08/04/opinion/invitation-to-a-dialogue-less-ego-in-architects.html?auth=login-email&login=email.

Evans, Peter. 2008. "Is an Alternative Globalization Possible?" *Politics & Society* 36 (2): 271–305. https://doi.org/10.1177/0032329208316570.

Fisher, Thomas. 2012. *Designing To Avoid Disaster: The Nature of Fracture-Critical Design*. 1st edition. New York: Routledge.

Frampton, Kenneth. 1983. "Prospects for a Critical Regionalism." *Perspecta* 20: 147–62. https://doi.org/10.2307/1567071.

Gautam Bhatia. 1994. *Laurie Baker: Life, Works, & Writings*. New Delhi: Penguin Books.

Harvey, David. 2009. *Social Justice and the City*. Revised edition. Athens: University of Georgia Press.

Mingle, Jonathan. 2015. *Fire and Ice: Soot, Solidarity, and Survival on the Roof of the World*. 1st edition. New York: St. Martin's Press.

Nussbaum, Bruce. 2010. "Is Humanitarian Design the New Imperialism?" Fast Company. July 6, 2010. www.fastcompany.com/1661859/is-humanitarian-design-the-new-imperialism.

Palleroni, Sergio. 2011. "Cultivating Resilience: The BaSiC Initiative." In *Beyond Shelter: Architecture and Human Dignity*, edited by Marie J. Aquilino, 222–33. New York: Metropolis Books.

Pearson, Jason. 2002. *University/Community Design Partnerships*. 1st edition. New York: Princeton Architectural Press.

Perkes, David. 2009. "A Useful Practice." *Journal of Architectural Education (1984–)* 62 (4): 64–71.

Rieff, David. 2003. *A Bed for the Night: Humanitarianism in Crisis*. Reprint edition. New York: Simon & Schuster.

Scott, James C. 1998. *Seeing Like a State: How Certain Schemes to Improve the Human Condition Have Failed*. New Haven: Yale University Press.

Wickersham, Jay. 2014. "Code of Context: The Uneasy Excitement of Global Practice." *Architecture Boston*, Winter 2014. www.archdaily.com/590300/code-of-context-the-uneasy-excitement-of-global-practice/.

6 Artificial glaciers

Artificial glaciers are engineered landforms that help high Himalayan communities divert, store, and manage irrigation water over the arc of a calendar year. By exploiting site conditions such as solar orientation, elevation, slope, and reliable freeze-thaw cycles, farmers form surface meltwater into large ice reservoirs. These are design interventions between 13,000 and 15,000 feet, massive infrastructural projects situated just above villages that act as an on-off tap for water management. Although artificial glaciers cannot solve the problem of shrinking glacial mass, they help farmers to make the most of this diminishing resource until a more sustainable solution can be found.

While the people of Ladakh commonly call these formations "artificial glaciers," the designs do not strictly fit within the accepted definition of a glacier, which is generally understood to be a naturally formed "slowly moving mass or river of ice formed by the accumulation and compaction of snow on mountains," stable over time (Oxford University Press 1989). Like a dam or a levee wall, they can be considered large-scale infrastructural works that are inextricably tied to context and site. In considering the artificial glaciers of Ladakh, one must understand that project designers have appropriated the term glacier to define any large-scale and relatively stable ice mass and as a colloquial term, "artificial glacier" is accepted and widely used in the region.

Artificial glaciers might be more accurately called engineered ice reservoirs, as they form large, layered pools of ice over the winter for spring and summer agricultural use. These large-scale landscape infrastructure projects efficiently store and manage glacial meltwater by exploiting annual freeze-thaw cycles and incorporating large storage reservoirs to contain glacial runoff. During periods of drought or when Ladakhi farmers otherwise wouldn't have access to meltwater (such as in the early spring months), artificial glacier pools melt, slowly releasing their ice stock in the form of irrigation water. The long, cascading pools are located between the high natural glaciers and the village croplands below, using gravity to move the water when farmers open regulating and diversion gates (Figure 6.1).

Figure 6.1 The artificial glacier pools above Igoo in the summer, showing the wall structure of each section.

These glaciers have been placed above nearly a dozen different villages in Ladakh, and range in size from a small number of pools at a single site to many dozens of reservoirs that extend over a mile in length. They are typically built using masonry walls with stone collected from the site and local labor, often through work parties organized by NGOs such as the Leh Nutrition Project. In enabling farmers to collect water that is cascading down through the watershed during non-agricultural seasons (fall and winter), and save it for later use (spring and summer), the artificial glacier effectively increases the overall water supply available to farmers (Vince 2009). Moreover, because these glacial pools serve the entire village and help to stretch the communal supply of water, they are collectively owned, managed, and operated by local village stakeholders.

Artificial glaciers present a low-tech form of geoengineering to address climate-induced water scarcity, using gravity and freeze-thaw cycles to control meltwater access and flow. These human-made stone and earthen reservoirs produce and hold ice masses as a form of water storage. Unlike many climate geoengineering experiments, these glaciers are physical structures that, once built, can be dismantled and effectively erased. In other words, these designs feature an "undo" option, should the experiment go awry.[1] Artificial glaciers present a design solution that connects the landscape to human development goals in clear and obvious ways, highlighting both the negative effects of climate change and making visible the promise of a human-centered design intervention.

In the arid landscapes of Ladakh, as natural glaciers recede and subsistence farming villages struggle to secure irrigation for their croplands, the geoengineering of artificial glaciers introduces a compelling interim solution to water scarcity (Grossman 2015). Pairing landscape architecture with hydraulic engineering to productively manage ecosystem services, this technique has already given rise to a host of new landforms across Ladakh. Today more than two dozen artificial glacier systems have been constructed in this region, as well as many hundreds of *zings*, or water reservoirs (Nelson 2009).

Background

Early examples of artificial glacier design thinking have been documented through oral history interviews with elderly Ladakhis as well as by looking for physical clues in current landscapes (Gladfelter 2018). While these early design interventions were not called artificial glaciers, they served a similar function to more modern systems by providing village farmers with additional meltwater reserves, trapped in winter months for use in the spring. These systems were built by village stakeholders, without outside monetary or labor support, and generally consisted of walls built in drainage areas to cause water to pool and then freeze.

The first system called an artificial glacier in Ladakh was designed by a local engineer, Chewang Norphel, in 1987 (Figure 6.2). His model was supported by the Rural Development Department (RDD) as a part of their Watershed Development Programme, and this work marks a shift away from traditional ad-hoc glacial pool construction to the current practice of reliance upon outside entities for funding and administrative support. The first officially recognized artificial glacier, above the village of Phuksey, still stands over a mile in length (Gladfelter 2018; Nelson 2009). Since then, Norphel has continued to champion this water harvesting strategy, overseeing the construction of at least 11 other artificial glaciers and many dozens of *zings*. He worked with the Leh Nutrition Project (LNP) for many years, and upon retirement, that NGO has continued to produce these projects. Despite the efforts of non-profit administrators and outside funders, these formations often demand extraordinary investment in terms of labor, materials, and engineering resources, both during the construction phase and throughout the life of the artificial glacier.

In the remote village landscapes of Ladakh, agricultural officers and hydraulic engineers have not definitively measured the amount of water annually coursing through existing waterways or newer artificial glacier systems. However, in recent years engineers have sought to quantify the amount of water held by artificial glaciers, and their findings demonstrate the productive capacity of this design intervention (Nüsser et al. 2019).

Figure 6.2 Chewang Norphel stands on a wall of the artificial glacier system he engineered above the village of Nang in 2015.

As is the case with any new climate-adaptive experiment, however, parameters for success and failure still need to be established. Without this data, Norphel and the Leh Nutrition Project rely primarily on visual analysis and experiential reports to adjust and improve the engineering of artificial glacier systems, effectively honing their designs with each built version.

In contrast to many of the radical climate geoengineering projects globally under consideration today, however, artificial glaciers have a long history of tested performance, with numerous applications and varying levels of investment. Villagers have tinkered with glaciers in remote Himalayan regions for centuries, using local labor and materials, and an intuitive understanding of site constraints (Tveiten 2007). Throughout the Himalayan mountain range, farmers have diverted glaciers since at least the nineteenth century, to convert ice masses into water for irrigation and to direct the flow of glacial meltwater. More recently, as glaciers diminish due to warming temperatures caused by climate change, some communities have adopted the practice of glacier grafting to seed new glaciers.[2] The idea of producing an ice reservoir for agricultural use is not new, and in the case of the artificial glaciers of Ladakh, their success and efficacy relies on many generations of local land use knowledge (Nüsser and Baghel 2016).

While farmers have historically tapped into natural glaciers for agricultural irrigation, artificial glaciers have also been incorporated into landscapes in more unconventional ways. Artificial glaciers may have been built as early as the twelfth century, as a military strategy used to block Genghis Khan's advance over high mountain passes (Khan 2005). In Russia, North Korea and Alaska, *naleds*, or constructed ice masses, serve as the seasonal ice bridges that facilitate winter movement in those areas. And in one astonishing example cited by *National Geographic*, the city of Ulan Bator, Mongolia, has developed a summer cooling strategy that incorporates these icy sheets (Inman 2011).

In Ladakh, however, the construction of an artificial glacier reflects a village's desire to reproduce or replace diminishing ice reservoirs in a context where natural glaciers once provided reliable irrigation services. *BLDG BLOG* author Geoff Manaugh calls these artificial glaciers a physical example of "vernacular hydrology writ large" (Manaugh 2010), and they do reference local building materials and techniques. Because each glacier is precisely engineered for a specific location, these landforms reflect a singular contextual strategy that will not be replicated elsewhere. The simple structures require steep, shaded topography, an elevation of at least 13,000 feet, and consistent freeze-thaw cycles to operate effectively (Norphel 2012). The artificial glaciers used for irrigation in Ladakh are necessarily rooted in place; they are site-specific installations embedded in the landscape.

As such, these landforms could be inspired and informed by the allied design disciplines, and offer valuable insights into performative landscapes of the future. The artificial glacier's precise siting, large-scale engineering demands, and integration into ancient village agricultural contexts suggest that it might benefit also from professional design thinking, despite the fact that designers have not contributed to glaciers built in Ladakh. Likewise, glacier sites have the potential to serve as more than simple ice reservoirs; they could benefit from design thinking around multifunctional landscapes or the appropriation of the water commons for more diverse uses. In this sense, artificial glaciers present a landscape type that could be similar to the *talaab*, or Indian water reservoir described by scholar Alpa Nawre, where due to "its integration in cultural life, the talaab builds a compelling ground for re-examining the design and application of contemporary 'water harvesting' techniques and water infrastructure" (Nawre 2013, 147). Both the talaab and the artificial glacier produce physical landscape enclosures for a village water commons, which could accommodate more diverse activities and services through intentional design programming.

How it works

Artificial glaciers trap meltwater from existing natural glaciers and seasonal snowpack to slow, store, and discharge this water supply in a more controlled way. These systems capture glacial meltwater in the fall and winter, directing it to a specific site through pipes and channels (in the case

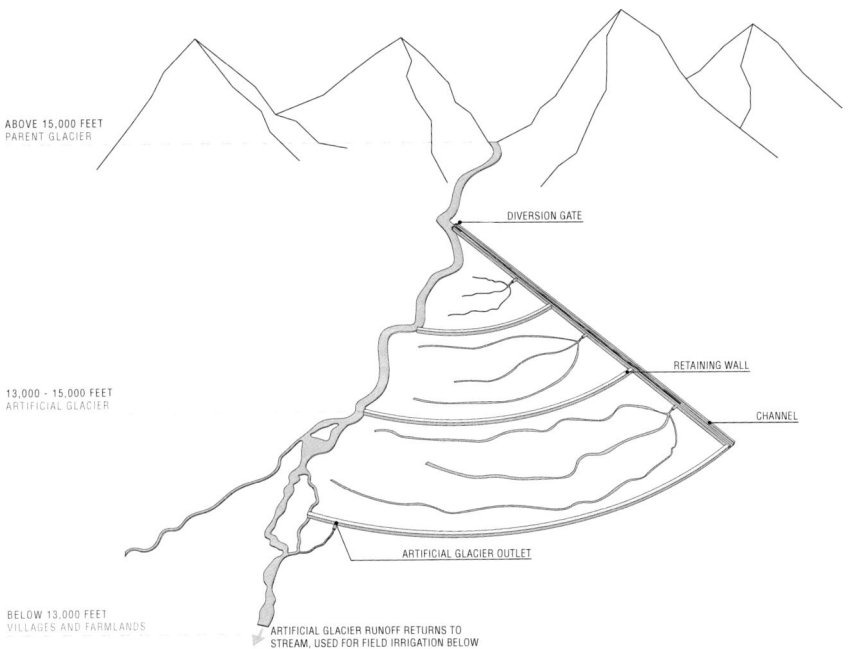

ABOVE 15,000 FEET
PARENT GLACIER

DIVERSION GATE

RETAINING WALL

13,000 - 15,000 FEET
ARTIFICIAL GLACIER

CHANNEL

ARTIFICIAL GLACIER OUTLET

BELOW 13,000 FEET
VILLAGES AND FARMLANDS

ARTIFICIAL GLACIER RUNOFF RETURNS TO
STREAM, USED FOR FIELD IRRIGATION BELOW

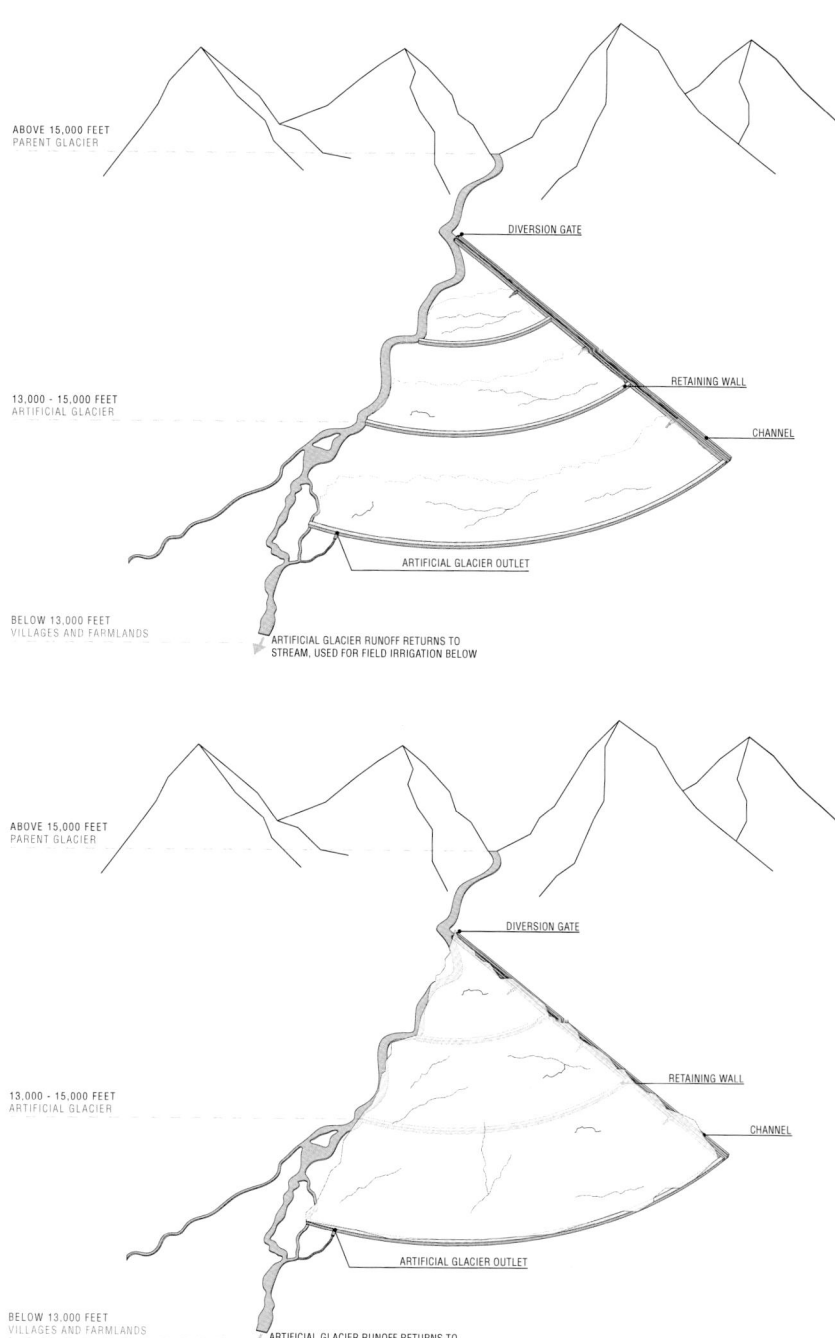

Figure 6.3 A diagram of a typical artificial glacier system.

of a diversion system) or via a stream bed, where it is then frozen and contained – stockpiled – for use in the spring planting months. Because these systems work across the arc of a four-season year, excess water from one season can be saved, in ice sheets, for following seasons. In Ladakh, artificial glaciers serve as the holding tanks and on/off taps for an otherwise erratic water supply (Figure 6.3).

These landscape interventions present as a low-tech engineering solution to water scarcity, using thoughtful placement (with regard to altitude, sun/shadow, slope, and relative position to villages and high natural glaciers) to allow nature to do the work. They can function at almost any scale, from a small pool to a large catch basin, and alone or in conjunction with other glacial pools. The construction of these landforms is relatively simple, incorporating local labor, building techniques, tools, and materials. Most diversion systems use land moving equipment to create a channel for water to flow in, and then pipes to spray water out across the pools. All artificial glaciers use local stone, usually dry stacked, and some employ concrete and steel mesh, particularly in small quantities to reinforce dry stacked stone walls (Figure 6.4). Because neither of these approaches represents a complicated system or new technology, the ice reservoirs can be built and maintained by local village stakeholders. Depending on size and location, the overall cost of one of these systems is between 6 and 20 thousand USD, roughly a quarter of the cost of a concrete reservoir for the same location (Vince 2009).

Although these formations necessarily exhibit unique, context-driven design features, the artificial glaciers in Ladakh share the same basic anatomy and engineering. Drawing from high mountain glaciers, water is directed toward the glacial pools. In a diversion system, wide stone channels divert water from the headworks of a natural glacier, slowing the water and directing it, through a series of underground pipes, to slowly discharge into a large stone reservoir. In an in-stream system, the same high glacier meltwater courses through a stream (rather than diversion canals) that has stone dams placed to slow and hold water as ice. In both systems, sheets of ice form, effectively holding fall and winter meltwater for the spring. When spring arrives, the ice bank, which is stored well below the glacier it came from (usually 3–4,000 feet lower) will melt sooner, and it can be used to jumpstart the new growing season.

Norphel's designs have evolved slightly over time, informed by both geography and accident. His projects, as well as those built or adjusted by the Leh Nutrition Project after he retired, include diversion systems, in stream systems, or combination systems. Regardless of shape, all of the projects function using the same core principles: they divert meltwater in fall and winter, slowly spreading this runoff into pools as icy sheets. In March or April, before high glaciers and snowpack begin to produce meltwater, the artificial glaciers will thaw. This early-season water cache is

Figure 6.4 From above, a single wall of the artificial glacier system above Igoo village.

accomplished without pumps or technology, by taking advantage of gravity and seasonal temperature changes, and choosing low-sloping, shaded sites between 13,000 and 15,000 feet. Most artificial glaciers have clean structural lines defined by massive stone walls and a series of cascading rectilinear pools: ice reservoirs extruded from the topography of adjacent mountains. The glaciers' straight lines and sheer scale conjure up images of earth-moving machinery when, in fact, laborers typically piece the walls together by hand.

Community engagement is a critical factor in the success of each artificial glacier, both in terms of leveraging resources to see the construction through to completion and in terms of assuring proper functioning, ongoing maintenance and critical repairs. Chewang Norphel notes that the first step in any new glacier design is to mobilize community participation: "Since the village communities are the main stakeholders and know the area and its dynamic thoroughly, the first step is to mobilize them and to hold intensive discussions with them" (Norphel 2012). Today the LNP hosts workshops with village stakeholders to establish design goals and work plans. This initial engagement prepares the villagers to contribute to the building and ongoing maintenance of the glaciers, while also providing the engineers with important information about sunrise, sunset, shaded areas, and water availability throughout the year. The early engagement efforts of NGOs not only assist in the construction phases, but also provide villagers with an understanding of the system so that they can participate in ongoing maintenance.

Of the many dozens of artificial glacier pools built by Chewang Norphel and the Leh Nutrition Project, not all have succeeded. According to Norphel, some of these glaciers have failed due to factors such as interpersonal conflict, abnormal flooding, mudslides, lack of maintenance, or improper calculation of temperature, elevation, and pool configuration. The three artificial glacier pools above the village of Stakmo were destroyed by a mudslide in 2010, for instance, reducing the storage capacity of each of the basins (Norphel 2012). Others, such as the glacier pools at Umla, and Muut, have gradually fallen apart due to lack of reinforcing steel mesh around the containment walls. In other cases, components of the systems, such as the glacier pools, had to be redesigned and rebuilt over years to produce desired outcomes (Gladfelter 2018). The challenge of maintaining artificial glaciers is an ongoing effort for village stakeholders (who occasionally hire it out) and represents difficult cold-weather work.

Villagers who live below these glaciers report varying levels of enthusiasm for the work, although in a survey by A. Kathleen Higgins of 40 stakeholders, feedback was overwhelmingly positive (Higgins 2012). Other scholars report that stakeholders have grappled with the task of maintaining these massive earthwork projects, especially if they are not producing a clear irrigation benefit (Gladfelter 2018). Scholar Sierra Gladfelter recently conducted a thorough survey of artificial glaciers in Ladakh in which she identified advantages and disadvantages of various systems, and this study found that low-tech, traditional systems often outperform the more highly engineered systems designed by Norphel (Gladfelter 2018). Despite the positive press about the artificial glaciers in magazines and newspapers, the real beneficial impact of these projects are still unclear: given the wide variety of locations and site-specific designs for these artificial glaciers, it is difficult to assess and compare their performance (Sudhalkar 2010). However, it is worth noting that since the official artificial glacier project's inception in 1987, new systems have been built at regular intervals with a variety of different techniques. Village stakeholders seek out NGOs and funding sources for new projects, such as the one just completed above the village of Likir in 2017, and the one slated for construction above the village of Saboo.

While the term artificial glacier may sound high-tech and energy-intensive, these landforms would be better characterized as low-tech infrastructural solutions. They have existed in some form for decades, employed by enterprising farming villages and time-tested in rugged high-altitude environments (Figures 6.5 and 6.6). Many villages in this arid landscape have a *zing*, or basic irrigation storage tank, which could be considered the progenitor to the artificial glacier. Both systems incorporate local masonry, exploit gravity, and capitalize on seasonal temperature shifts.

Figure 6.5 The artificial glacier above the village of Phuksey is partially melted.

Figure 6.6 Above Phuksey village, the artificial glacier holds snow and ice well into mid-summer.

Cost and production

The financial model for the development and management of artificial glaciers has undergone a number of changes over the course of this product design. Early artificial glaciers were primarily funded by non-profit partners and created with labor from villagers themselves, featuring some combination of DIY initiative on the part of village stakeholders, and the addition of external resources (such as monetary or material donations) for imported components. In one unusual example, an American group with educational ties to the area, called *The Institute for Village Studies*, worked directly with Chewang Norphel to engineer and oversee the building of three artificial glacier pools above the village of Stongde in the Zanskar Valley. Today the Leh Nutrition Project works with a variety of philanthropic organizations to build and restore artificial glaciers. The largest donor for these LNP-sponsored projects continues to be the TATA Trusts. However, efforts in recent years have also been funded by the Save the Children Foundation, the Watershed Development Programme, and the Army goodwill program called Operation Sadbhavana.

Benefits

While artificial glaciers will not add water to the high mountain landscapes, they help to maximize the use of existing water resources by storing and relocating glacial meltwater for irrigation. These glaciers effectively extend the growing season, by making the previous year's stored meltwater available to farmlands up to two months before meltwater would otherwise arrive. Timing is key: if farmers have access to water in April, rather than June, they benefit from a longer and more productive summer growing season. According to scholar A. Kathleen Higgins, this is particularly relevant for Himalayan farming, where agricultural cycles have become decoupled from the climate due to global warming (Higgins 2012). In this environment, water availability has increasingly moved out of sync with agricultural needs. In the context of climate change, where temperatures in the Himalaya have increased by an average of two degrees Celsius, "The ground now thaws by April, a full month earlier than two decades ago, but water still comes in May or June, limiting farmers to approximately four months of temperatures warm enough to grow crops" (Higgins 2012, 7). In addition to supplying water to farmers to extend the growing season, artificial glaciers can help to direct water to specific areas, conserve the supply of meltwater, and may even recharge the local aquifers of villages.

Limitations

In Ladakh, artificial glaciers suspend water (as ice) that would otherwise feed the headwaters of the Indus River. While it is unlikely that the more than dozen artificial glaciers in this region would substantially impact the flow of such an enormous river, there could be a reduction of flow downstream if this design solution were to scale up across the region. As long as artificial glaciers are built above villages that already trap meltwater for irrigation, they only help to improve the efficacy of those extant systems. Overall, the ecological impact of these design interventions is both minimal and site-specific, primarily causing water to be held longer on site and then diverted to nearby fields for agricultural use.

   While artificial glaciers can be understood as beneficial structures for farmers grappling with chronic drought, the temporary nature of this design solution requires one to consider this work as a bridge between current instability and a future that might become more sustainable.[3] Artificial glaciers do not increase the overall supply of water available to a farming community, and they rely upon annual snowmelt and established glaciers at higher elevations in order to function (Norphel 2012). Due to receding natural glaciers, artificial glaciers also have a built-in sunset date; thus they present as a short-term mitigation rather than a permanent adaptation. People searching for solutions to the problem of receding glaciers may be hooked by the idea of a human-made system, and indeed the concept of artificial glaciers offers a tempting kernel of hope for environmental adaptation. But artificial glaciers must draw from extant glaciers, and while they efficiently use that water source, they do nothing to replenish glacial mass.

Moving forward

Proponents of artificial glaciers note that these landscape interventions could scale up, improving water access across and beyond north India. Chewang Norphel has plans for artificial glaciers in at least one hundred villages, extending beyond Ladakh into Pakistan, Kazakhstan, and Kyrgyzstan (Shrager 2008). Ice stupas and icefall glaciers have built upon the design idea of the artificial glacier to test new adaptation strategies. These and future glacier-making initiatives would likely draw from the lean field-based operations of the past: referencing unique site and topographic features, soliciting useful oral histories from farmers, and minimally sculpting extant landscapes. Today, the Leh Nutrition Project is in the process of assisting the village of Saboo with a new artificial glacier system, and additional glacial prototypes are underway.

Despite these interventions, climate change is eroding the glaciers in Ladakh with alarming speed and, in many cases, irrevocable outcomes. Artificial glaciers rely on natural glaciers at even higher elevations, and the annual snows, precipitation, and cold winters that sustain them, to exist. Indeed, if the highest glaciers in this region were to disappear completely, the artificial system would also cease to exist. This regional connectedness is perhaps the largest design factor for the success or failure of these landforms, and is a reminder that adaptation efforts do not necessarily obviate the need for climate change mitigation.

While artificial glaciers in northern India may stem the problem of chronic drought in the short-term, they fail to address the larger crisis of a warming planet. For the receding glaciers in Ladakh, as around the world, only additional precipitation can replenish the mountains' great stock of ice. Therefore artificial glaciers cannot replace diminishing natural glaciers, or even protect arid landscapes from eventual crushing drought – but they could improve ecological functioning in the interim. These irrigation systems ease the immediate stranglehold caused by water shortages, and empower farmers to conserve, allocate, and store water for their seasonal needs. While the use of artificial glaciers in Ladakh may only provide a viable approach to water scarcity up to a certain point, as long as they mitigate the effects of climate change locally, they provide a much-needed intermediate solution for struggling farmers (Nüsser, Schmidt, and Dame 2012). As such, they may buy critical time for drought-impacted mountain villages, until a more sustainable solution can be developed.

In this sense, a useful critique of Ladakh's artificial glaciers may ultimately hinge on the time horizons one assigns to this work. There is a real disciplinary tension between the long-term vision and 'ultimate' solutions that have historically been preoccupations of architects and engineers, and the unfinished or transformative schemes that reject the idea of a single, fixed solution in favor of a more flexible design agenda. In the case of Ladakh's artificial glaciers, these two approaches may find common ground in a long-term transformative solution, in which the temporary needs of villagers can be satisfied while working towards more sustainable systems.

In the context of the ever-expanding design disciplines, artificial glaciers stand out as just one example of the type of climate change project that would benefit from greater engagement with design professionals. At the heart of this disciplinary refocusing is an acceptance of the unknown, recognizing that

> Landscape architects and urbanists can help reverse the process (of environmental deterioration), cognizant that even with our best intentions, the landscape we create may yield unpredictable results, and that the aspect of 'change' is the underlying factor in everything we do.
> (Hung 2013, 14)

Indeed, this change is coming to characterize the profession and, hopefully, it will open new doors for designers in the process. As climate change influences landscapes in new ways, the vision and optimism of design thinking could offer a useful and important form of service. However, while large-scale infrastructural projects such as artificial glaciers appear to mediate the negative effects of climate change in the short term, these solutions may only provide temporary relief from the larger crisis of a warming planet.

Conclusion

Despite widespread popular interest in artificial glaciers, very little scholarly research has been published on this topic, a tiny fraction of which would qualify as design discourse (Vince 2009). One of the few design writers addressing this topic is *BLDG BLOG*'s Geoff Manaugh, who connects these structures to the broader issue of climate change adaptation in claiming that these villagers "have learned to reorganize their region's existing snowpack so as to make it thermally self-sustaining" (Manaugh 2010). The formal and spatial qualities of these glaciers reflect performative and logistical concerns; they are utilitarian and understated ice formations that derive their physical logic more from gravity and topography than experience or formal composition (Figure 6.7).

If artificial glaciers have been designed with an engineer's imperative, the striking lines, reflective pools, and rooted, low-slung elegance that characterizes each site suggests that this has been a valuable formal generator. Like a dam or levee, these projects can be massive; one glacier measured is 1,000 feet long, 150 feet across, with an average depth of 4 feet (Than 2012). The minimalist aesthetic of these formations derives from earth-formed walls that blend into the surrounding topography despite their clear organization and solid, severe lines. These projects incorporate materials gathered from the immediate site (earth and rock), and so while the monolithic structures are clearly visible from Google Earth images

Figure 6.7 An artificial glacier in winter becomes an ice field.

taken from 30,000 feet, on the ground they almost disappear completely into the surrounding landscape. Similar to contemporary landscape projects in Europe, these constructed landscapes support ecological functioning while introducing striking new geometries. Formally, artificial glaciers appear to be elegant yet monumental landforms, strategic incisions into a dry and monochromatic landscape.

While Ladakh's artificial glaciers stand out as an exciting precedent for a new genre of climate-adaptive landscapes, they lack the formal design and planning that might help these spaces capitalize on synergistic relationships. For example, designers might look beyond the primary use of artificial glaciers (as water storage spaces) to identify secondary uses around recreation, tourism and cultural heritage. While the shape and structure of these glacier pools express a technical approach to the problem of water scarcity, the stone walls and even the ice contained inside the glaciers have become a visible part of the landscape with very little experiential value. In purely functional terms, Ladakh's winter ice skating and summer trekking – both important social and economic pursuits for villagers – could be built onto and around the artificial glaciers. Designers could bring value to the development of artificial glaciers, just as in other engineered environments, they have influenced the nature and the form of utilitarian projects by sculpting land in experiential and expressive ways.

The artificial glaciers of Ladakh stand out as a design solution that enables villages to make do with fewer resources. In this sense, they have become an example of engineering applied in the service of better resource management, and in which case the outcome just happens to be a wonderfully expressive new landscape feature. As artificial glaciers continue to be built and improved upon in Ladakhi villages, perhaps new applications for utility and human access will also emerge. In the interim, artificial glaciers demonstrate a working model for the value of design interventions in small mountain communities. They also have begun to demonstrate the potential for foreign and national philanthropic partnerships, and associated applications for tourism. In so doing, the artificial glacier project provides a model for climate-adaptive design thinking that could be appropriated elsewhere in the world.

Notes

1 The Intergovernmental Panel on Climate Change (IPCC) concluded in 2007 that geoengineering options for climate change "remained largely speculative and unproven" (Intergovernmental Panel on Climate Change 2007). People worry that without an undo option, geoengineering experiments could go very wrong and leave countries with little recourse for improvement.
2 In Pakistan glacier growing is a longstanding practice, but since 2005, the Aga Khan Rural Support Programme in Skardu, Pakistan has been monitoring 18 grafted glaciers in Pakistan for their impact on agriculture (Khan, 2005).
3 The lifecycle of individual artificial glaciers varies according to the size of the natural glacier that feeds the artificial glacier, anticipated rainfall or snowfall, and overall temperature increase in the future. Projections for the sunset date would mirror those of natural glaciers, with erosion of the glaciers occurring in the next 10–25 years. While 25 years may appear to be a very short design intervention in the long term, such an affordable and efficient solution could prove reasonable in the interim.

References

Gladfelter, Sierra. 2018. "Ladakh's Artificial Glaciers, Ice Stupas, and Human-Made Ice Reserves." Fulbright-Nehru Student Research Report. United States-India Educational Foundation.

Grossman, Daniel. 2015. "As Himalayan Glaciers Melt, Two Towns Face the Fallout." Yale E360. March 24, 2015. https://e360.yale.edu/features/as_himalayan_glaciers_melt_two_towns_face_the_fallout.

Higgins, A. Kathleen. 2012. "Artificial Glaciers and Ice-Harvesting in Ladakh, India as an Adaptation to a Changing Climate." New Haven, Ct: Yale School of Forestry and Environmental Studies.

Hung, Ying-Yu. 2013. "Landscape Infrastructure: Systems of Contingency, Flexibility, and Adaptability." In *Landscape Infrastructure: Case Studies by SWA*, edited by Ying-Yu Hung and Gerdo Aquino, 14–19. Basel: Birkhauser.

Inman, Mason. 2011. " 'Ice Shield' Experiment Aims to Cool Mongolian City." *National Geographic*, December 2, 2011. http://news.nationalgeographic.com/news/2011/11/111207-global-warming-geoengineering-mongolia-ice-science-environment/.

Intergovernmental Panel on Climate Change. 2007. *Climate Change 2007: Mitigation of Climate Change: Contribution of Working Group III to the Fourth Assessment Report of the Intergovernmental Panel on Climate Change.* Cambridge ; New York: Cambridge University Press.

Khan, S. 2005. "Glacier Grafting." The Aga Khan Rural Support Programme. www.akdn.org/download/2011_akdn.pdf.

Manaugh, Geoff. 2010. "Artificial Glaciers 101." *BLDG BLOG* (blog). February 26, 2010. http://bldgblog.blogspot.com/search?q=artificial+glacier.

Nawre, Alpa. 2013. "Talaab in India Multifunctional Landscapes as Laminates." *Landscape Journal* 32 (2): 137–50. https://doi.org/10.3368/lj.32.2.137.

Nelson, Dean. 2009. "Indian Engineer 'builds' New Glaciers to Stop Global Warming." *The Telegraph*, October 28, 2009, sec. Home. www.telegraph.co.uk/earth/environment/globalwarming/6449982/Indian-engineer-builds-new-glaciers-to-stop-global-warming.html.

Norphel, Chewang. 2012. "'Artificial Glacier: A High Altitude Cold Desert Water Conservation Technique.'" In *In Defense of Liberty Conference Proceedings*, 3. New Delhi, India: Leh Nutrition Project. http://indefenceofliberty.org/story/4343/4495/Abstracts-of-papers-Session-3-April-9-10Climate-Change-Understanding-Himalayan-Ecology.

Nüsser, Marcus, and Ravi Baghel. 2016. "Local Knowledge and Global Concerns: Artificial Glaciers as a Focus of Environmental Knowledge and Development Interventions." In *Ethnic and Cultural Dimensions of Knowledge*, edited by Peter Meusburger, Tim Freytag, and Laura Suarsana, 191–209. Knowledge and Space. Cham: Springer International Publishing. https://doi.org/10.1007/978-3-319-21900-4_9.

Nüsser, Marcus, Juliane Dame, Benjamin Kraus, Ravi Baghel, and Susanne Schmidt. 2019. "Socio-Hydrology of 'Artificial Glaciers' in Ladakh, India: Assessing Adaptive Strategies in a Changing Cryosphere." *Regional Environmental Change* 19 (5): 1327–37. https://doi.org/10.1007/s10113-018-1372-0.

Nüsser, Marcus, Susanne Schmidt, and Juliane Dame. 2012. "Irrigation and Development in the Upper Indus Basin." *Mountain Research & Development* 32 (1): 51–61. https://doi.org/10.1659/MRD-JOURNAL-D-11-00091.1.

Oxford University Press. 1989. *The Oxford English Dictionary.* 2nd ed. Oxford: Oxford; New York: Clarendon Press; Oxford University Press.

Shrager, Heidi. 2008. "'Ice Man' vs. 'Global Warming.'" *TIME*, February 25, 2008. http://content.time.com/time/world/article/0,8599,1717149,00.html.

Sudhalkar, Amruta Anand. 2010. "Adaptation to Water Scarcity in Glacier-Dependent Towns of the Indian Himalayas : Impacts, Adaptive Responses, Barriers, and Solutions." Thesis (MCP), Massachusetts Institute of Technology, Dept. of Urban Studies and Planning.

Than, Ker. 2012. "'Artificial Glaciers Water Crops in Indian Highlands.'" *National Geographic*, 2012. http://news.nationalgeographic.com/news/2012/02/120214-artificial-glaciers-water-crops-in-indian-highlands/?rptregcta=reg_free_np&rptreg campaign=20131016_rw_membership_r1p_us_se_w#finished.

Tveiten, Ingvar Nørstegård. 2007. "Glacier Growing – A Local Response to Water Scarcity in Baltistan and Gilgit, Pakistan." Thesis, Norway: Norwegian University of Life Sciences.

Vince, Gaia. 2009. "Glacier Man." *Science* 326 (5953): 659–61.

7 Ice stupas

The ice stupa is a close cousin to the artificial glacier, as both projects produce a physical cache of water, frozen in the winter months, for farming use in the following spring. Compared to other climate-adaptive design efforts underway in Ladakh, it is also one of the most multifaceted: project designers intentionally overlay ambitious environmental, economic, social, cultural, and religious goals in the production of a single landscape feature. During the course of the five-year period from 2013 to 2018, ice stupa prototypes built in the village of Phyang have demonstrated both benefits (primarily in terms of irrigation production) and challenges (associated with water access and ownership). Since then, dozens of additional ice stupas have been built in other villages in Ladakh, making a case for a more widespread application of the design intervention.

In order to understand how the ice stupa works, it is helpful to visualize water flowing through the topography of the mountain range, which harbors high glaciers and snowfields. Ladakhi villages already capture meltwater for their use, as it moves towards the headwaters of the Indus River. As winter water courses through this landscape via mountain streams, a portion of it can be diverted to form an ice stupa, which will later be tapped for agricultural use during the growing season. The ice stupa is a freestanding pyramidal stock of ice, formed in the drainage below glaciers yet above villages, that melts in the spring to provide farmers with a secondary source of water.

The Ice Stupa Team was formed in 2013 by engineer, educator, and entrepreneur Sonam Wangchuk, and is composed of volunteers, staff, consultants, and student interns. Although Sonam Wangchuk leads this work, the entire Ice Stupa Team could be considered project designers, and the team works together to solve problems and explore new ideas. While prototypes were built annually since the winter of 2013–2014, until 2018 there was limited adoption across Ladakh, which hampered widespread testing of the project's structure and efficacy. In 2019 and 2020, several dozen ice stupas have been built across Ladakh, and so there is a growing body of these projects with which to collect data.

The Ice Stupa Team initially announced their temporal landscape design as an environmental enhancement, but throughout the life of the

Ice Stupa Project they have intentionally layered over additional economic, social, religious, and cultural considerations. As a small-scale design for water management, the ice stupa provides a replicable model for farmers and village communities. Beyond this explicit utility in Ladakh (if it becomes widely adopted), the project more broadly serves to demonstrate that climate-adaptive design is an unfolding practice, that layers of complexity may be built into all manner of environmental interventions. In this sense, the Ice Stupa Project provides an instructive case study for broader climate-adaptive design thinking, and a working reminder of the value, and perils, of conceiving of projects in multiple dimensions.

Background

The ice stupa assumes a commanding and conspicuous form in the expansive desert landscapes of Ladakh, and once formed it also appears to occupy a celebrated status in the collective village consciousness (Figure 7.1). In previous years, ice stupas have grown as high as 64 feet tall and their sites have been used for various cultural events, Buddhist prayer, and tourism (Strochlic 2017). The iconic form is easily recognizable due to its incongruous appearance amid Ladakh's desert landscapes of sand and rock. And as a design intervention, it has garnered considerable attention in local, Indian, and international press (Saxena 2015; Strochlic 2017).

Sonam Wangchuk, a Ladakhi engineer, educator, and entrepreneur, developed the initial ice stupa design idea in 2012–2013. Since then he has led the Ice Stupa Project, creating an Ice Stupa Team to erect prototypes, while also lecturing, fundraising, and negotiating with various stakeholders to complete annual builds in Phyang Village and beyond. According to Wangchuk, the idea is simply about putting to use "the fresh snow and ice in the mountains that melt even in winter and goes (to) waste," by creating a vehicle for it to be "frozen and stored until spring when farmers need (it) the most" (Wangchuk 2015). Indeed, this design harnesses and holds winter meltwater in place for use during the next growing season as an ice stupa.

The ice stupa is a direct descendant of the artificial glacier and borrows many design ideas from that structure. The two designs enable water stockpiling, and functionally provide farmers more control over glacial meltwater reserves. The ice stupa diverts flowing water that comes from a mountain stream – drawn from higher glaciers and snowfields – during freezing winter months. Using a network of gravity-fed pipes, the water is sprayed up into the air, forming the crown of the pyramid: layer by layer, an ice tower forms (Figure 7.2). The result is a conical block of ice that can endure as a frozen mass for several months, thereby storing water that would have been unusable for village agriculture in the winter for use in the following spring and summer seasons.

Figure 7.1 An ice stupa sprays water from the crown during winter months.

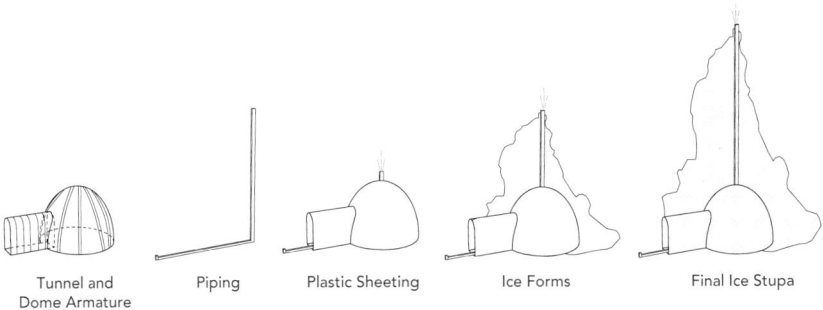

Figure 7.2 A diagram of a typical ice stupa system.

Rather than carving into existing mountain topography, in the manner of an artificial glacier, ice stupas grow up in increments into freestanding ice towers. Because the pipe that reaches up to the crown of the stupa will eventually ice over as the structure grows, it must be moved at intervals to accommodate the stupa's increasing height. This on-site management is a cold-weather, labor-intensive effort for builders during the winter months. Team members actively create ice stupa structures each year, moving a central pipe, by hand, to the top of stupa's crown as it grows.

In Phyang, Ladakh, the Ice Stupa Team managed this building process from 2013–2018 by camping out at the site during winter months. While their project was located on land just below the Phyang Monastery, it occupied a site somewhat distant from village households, and thus had to be overseen from a temporary camp. During the summer months during this time period, team members placed and tended underground pipes throughout upper regions of Phyang, in order to divert water from the main village stream to their site in the winter. The team initially placed 10 cm (4") pipes to draw meltwater to the project site, but in an effort to create ever larger structures (and to increase the water held for irrigation use) the network was completely replaced with 20 cm (8") and then 28 cm (11") pipes in 2017 (The Ice Stupa Project 2019; Gladfelter 2018).

According to their own literature, the Ice Stupa Team's first two-story prototype held 150,000 liters of winter stream water that would otherwise have coursed through the drainage in the winter resulting in water that would not available for farming use during the growing season (Wangchuk 2015). Sonam Wangchuk estimates that ice stupas could contain up to 10 percent of the water coursing through the Phyang desert, while an additional 10 percent would be used by villages, and the remaining 80 percent would eventually join the Indus River (Parvaiz 2018). Although they were conceived as an adaptive aid to farmers, the Phyang prototypes melted in a barren desert location with newly planted trees, rather than in a more established space for land cultivation.

While the Ladakhi ice stupas are a newer design concept than artificial glaciers and thus have fewer years of supporting data, they draw upon a similar set of engineering principles. Unlike the glaciers, which require large, shaded sites at specific altitudes, ice stupas can be built on smaller land parcels above an agricultural holding, at any location that freezes. The stupas can be built on almost any surface, with any solar aspect, as long as a gravity-fed stream can be diverted to feed the site from above. The ice stupa shape also has an advantage to the artificial glacier: There is less surface area exposed to the wind and sun than there would be with a flattened ice field, which reduces losses due to sublimation. While ice stupas may store a fraction of the water held in the large reservoirs of artificial glaciers, their relatively small scale could facilitate more extensive deployment throughout Ladakh.

As a design intervention, the ice stupa is an adaptive response well suited to the Ladakhi social, economic, and environmental context. It is inherently tailored to each site, and operates within the region's broader weather patterns and seasonal changes. It is an affordable technological adaptation using primarily gravity and human labor, requiring only a few imported components, such as pipes and fittings. Prototypes have accommodated the social, cultural, and religious interests of Phyang village and monastery. And the design fundamentally addresses the problem of water scarcity: By collecting water that naturally courses through the watershed during late fall and winter, and holding it for use in spring, the ice stupas make efficient use of an otherwise untapped water resource for agricultural use.

Religious, social, and cultural connection

In comparing the shape of these ice caches with Ladakhi reliquary mounds, the Ice Stupa Team chose a name that reflects their intention to associate innovative design thinking with the region's Buddhist religious fabric. The conical shape formed, and the color of ice highlighted by the brown desert context, closely resemble the ubiquitous white stone chortens that have, over the centuries, dotted the Ladakhi landscape (Devers 2017). This Buddhist connection has also been solidified with physical acts and partnerships. The Project's first series of prototypes were tested on a pilot site that was provided by the Phyang Monastery. Regional monastic leadership (such as His Holiness Drikung Kyabgon Chetsang Rinpoche) have further helped to integrate the work into normative religious practices and celebrations, by hosting events at the site and blessing each of the stupas annually.

As a strategy, the technique of intertwining Buddhist religious values with environmental activism is not new (Darlington 2012), but this approach serves as a powerful form of support in the cultural and social context of Ladakh. Here many Buddhist farming families look to local monastery leadership for guidance and direction, especially with regard to annual farming decisions and collective land stewardship practices. In declaring its support of the Ice Stupa Project, the Phyang Monastery has effectively sanctioned an otherwise unorthodox farming practice within the region's staid agricultural traditions.

Buddhism is only one of the religions represented in Ladakh, and the overt religious branding of the ice stupa may not capture the interest of non-Buddhist stakeholders. This concern is one example of the way in which combining religious interests and environmental efforts may produce areas of tension in practice. Regardless, as a project that may otherwise have stood out for its radical appearance and new environmental goals, the ice stupa has instead been brought into agreement with some of the region's longstanding religious, social, and cultural values.

Religion is a powerful driver of influence and activism in Ladakh, but so too are social and cultural practices (Gagné 2016). In recent years, there has been extraordinary pressure on the Ladakhi people to conform to the standards of the Indian government (Mingle 2003), to distinguish their land as a region separate from the adjacent nations of China and Pakistan (Nüsser and Baghel 2016; Gagné 2017; Rizvi 1998), and to retain cultural identity in an ever-globalizing planet. While the unveiling of a quirky new technique for farming might appear to be at odds with Ladakhi distinctiveness, the team's use of the ice stupa site for traditional Ladakhi dances and festivals, Buddhist *puja* (prayer) and other forms of social interaction has instead helped to ground it within the collective Ladakhi consciousness.

Moreover, such religious, social, and cultural layering suggests that the Ice Stupa Team has adroitly forged a network of alliances that may ultimately support the project's success. Although the Ice Stupa Project only celebrates one religious sect among many in the region, this alliance galvanizes support for adoption among the large Ladakhi and immigrant Tibetan Buddhist population. While it may be the case that "a technological solution to environmental change can rarely stand alone, but will always become enmeshed with local concerns and beliefs" (Gagné, Rasmussen, and Orlove 2014, 804), this project may also serve as an example of an intentional and explicit effort on the part of project designers to weave their work more deeply into social webs.

In the case of the ice stupa, this is just one of the creative and unconventional partnerships that not only organizes the project's framing and its rhetoric, but also suggests new possibilities for other types of climate-adaptive design co-optation. The ice stupa begins by borrowing environmental engineering from the artificial glacier design, but also moves beyond this antecedent to pair water husbandry with a number of the region's extant social, cultural, and religious values. In so doing, this model acknowledges the co-benefits of interdisciplinary environmental stewardship projects and provides a template that could be applied to integrated environmental action in other parts of the globe.

Economic opportunities

Grey literature sources routinely cite potential economic benefits that could be provided by ice stupa sites, beyond the addition of water to support agricultural irrigation ("Ice Towers in the Desert – Rolex Awards" 2016; Strochlic 2017). A central goal for the project in Phyang has been to use the ice stupa prototypes to expand tourism activity, including creating a network of village homestays. This objective has already been born out, as the sensation of the ice stupas have attracted "thousands of visitors" to Phyang village, and the Ice Stupa Team serves as an intermediary to connect travelers with homestays (Parvaiz 2018; Verma 2018). A more ambitious project goal is to one day create an ice hotel, either within or around an ice stupa formation (Wangchuk 2018). Below Phyang Village, the initial stupa prototype site has been used to irrigate a field of over 5,000 thriving saplings on a site that was once a desert, demonstrating the value of the additional water reserve. Rather than directing this irrigation toward existing farming areas, the Ice Stupa Team has used the new irrigation to demonstrate, in vivid greenery, the capabilities of the stupas.

A case for complexity

Since its inception in 2013, the Ice Stupa Team has actively engaged other individuals, including farmers, monastic leadership, and students, and visiting volunteers and researchers in order to foster collaboration, funding, and other types of support. This practice has had the effect of forging common ground between groups with radically different interests, and it has resulted in a project that weaves strands of disparate agendas in a single coherent scheme.

While the project is such a low-tech and small-scale design that it could be essentially free to produce, it has been supported via crowd funding and may be supported with money from various awards. The ice stupa has been featured in articles by *National Geographic* and the *New Yorker*, and Sonam Wangchuk has been awarded the prestigious international Rolex Prize, in part for this work.

Through a competition, new stupas were built in at least ten villages in the 2018–2019 winter, some 20 stupas across the 2019–2020 winter, and prototypical projects have also been designed in Nepal and Switzerland (Kolbert 2019). The project, which has been widely in a variety of popular websites, magazines and newspapers, has captured the imagination of people who otherwise might not have a connection to Ladakh. This outreach has enabled the Ice Stupa Project to grow in scale and scope, to impact individuals with interests beyond the project's geographic watershed, and, ultimately, to create an environmental enhancement that also satisfies the more prosaic needs of stakeholders.

Limitations

A number of limitations can be observed in the Ice Stupa Project, particularly with regard to its lack of widespread adoption, appropriated site for pilots in Phyang, and early challenges about access and equity. Perhaps the most serious charge against the Ice Stupa Project to date is that it has caused negative impacts for downstream stakeholder groups (Parvaiz 2018), and it is one that threatens the viability of the project in its current form.

The lack of incremental technological adoption highlights a major design hurdle for the ice stupa, and one that also reveals the difficulty of introducing new innovation models in order to address climate change. According to Everett Rogers' *Diffusion of Innovations* (Rogers 2003), new ideas take time to work through social systems to gain widespread adoption, and without a critical mass of followers, an idea will not be sustained. Because the ice stupa is a labor-intensive project that requires regular winter oversight, the effort that it requires in terms of tending pipes during the winter months may ultimately represent an unreasonable burden for area farmers. In the past, Sonam Wangchuk has expressed a desire to automate the building process, primarily by using sensors and other technological fixes. However, these modes have yet to be developed and even if they were available they would almost certainly increase the overall system cost.

In order to promote this vision, the Ice Stupa Team, and its primary patron, the Phyang Monastery, have intentionally used "Go Green" rhetoric. Early ice stupas effectively demonstrated this capacity to green a desert by rerouting water from the main tributary to feed a field of newly planted poplar and willow trees (Figure 7.3). While this field demonstration helps to "promote the project amongst foreign donors," it also could be in violation of the legal land use designation for the area, and contrary to the stated project goals, does not operate in the service of area farmers (Sharma and Bharat 2017, 24).

Beyond technical complications, the ice stupa prototypes built in Phyang raise questions about equity, accessibility, and the right to water resources that have traditionally been held as a form of the village commons (Angchok, Stobdan, and Singh 2014; Gilmartin 2003). The possibility for an ice stupa formation to create downstream losers by diverting water is not just hypothetical, but now confirmed. In an effort to demonstrate that the ice stupa can create a massive ice reservoir for agricultural use, and to simultaneously green a section of desert, the Ice Stupa Team has diverted a significant portion of the Phyang river drainage. According to residents downstream, all of the water in Phyang Nallah (the main village stream) is diverted to the ice stupa during winter months, without returning that water volume back into the same watershed.

Just downstream from Phyang village is Phey village, where residents rely on winter meltwater to recharge their aquifer, to support the needs of

Figure 7.3 A field of poplar and willow trees has been planted below the Phyang ice stupa site.

local flora and fauna, and to wash their cars and household goods. Here residents argue that winter water is not "wasted," as Wangchuk once asserted, but both anticipated and used. When the pipes were enlarged in 2017 to 11 inches, Phey residents said that they noticed a significant loss of their winter water in the main tributary, because it was all diverted to the stupas, whereas typically some water would be visible moving through the main stream channel (Parvaiz 2018). In the Ice Stupa Project's third year, over the winter of 2016–2017, stakeholders became engaged in a bitter public water rights debate. The Ice Stupa Project was criticized for robbing downstream landowners (in Phey village and also lower Phyang) of their right to water from the primary drainage (Maqbool 2019; Parvaiz 2018). Indeed, the Ice Stupa Team had diverted water from this main waterway for their use in building the stupa, blocking off access to those below (Sharma and Bharat 2017, 17).

In the spring of 2017, Sharma and Bharat conducted a study of water rights impacting the Phyang/Phey conflict, in which they discovered that Phyang held "inordinant power" over the lower village of Phey, due to the cursory coverage of legal rights detailed by a settlement officer in 1908. This survey was built into Ladakh's transition into British rule, and then was absorbed by the Indian government under independence. In the winter of 2018–2019, the Ice Stupa Team announced a competition to inspire stupa builds in other villages, with cash prizes for winners. Eleven villages took up this challenge, and these efforts represent the most widespread adoption of the ice stupa technique across Ladakh to date (Kolbert 2019). This pivot – away from a single contested site in Phyang to instead build stupas on multiple uncontested (and celebrated) sites elsewhere in Ladakh – does more than just sidestep critics, but fundamentally helps to strengthen project goals of broader visibility, adoption, and dissemination.

A critical consideration for the evaluation of potential climate-adaptive design projects is its impact on other systems, especially if it might create 'downstream losers' (Magnan 2014). One of the central principles of the Ice Stupa Project, and indeed of artificial glaciers as well, is the assertion that an ice reservoir is built using "waste" water during cold winter months. If a community actually does rely upon meltwater in those winter months, even for such inconspicuous uses as groundwater recharge or broad ecosystem functioning, then the water diverted to reservoirs could negatively impact whole system health.

Moving forward

It is the entwining of a full ambit of project factors – social, cultural, religious, economic and environmental – that sets the Ice Stupa Project apart from other water management initiatives. Instead of setting the single goal to alter an environmental condition, the Ice Stupa Project addresses multiple functions and types of stakeholder interests. The project is community-based, with local advocates, volunteers, and management. It incorporates imported pipe materials, but is fine-tuned to the specific climate and context of the region. The ice stupa incorporates references to the cultural and religious design language of Ladakh, and it overlays an extensive social network for fundraising. Such a multi-faceted approach to environmental improvement has much to teach the design disciplines about harnessing complexity in the service of climate-adaptive works.

The Ice Stupa Project also fosters ambitions beyond better environmental stewardship or more efficient water provisioning. The use of a stupa for a hotel, or prayer site, or other types of cultural programming could enrich the limited functioning of a desert ice tower (Figure 7.4). It is worth remembering that the original engineering and expressed intention of the ice stupa design was to produce a stock of water for seasonal agricultural use. In this capacity, the ice stupa provides an elaborate, unorthodox, and labor-intensive example of the lengths that Ladakhis may turn to in order to catch and retain a secondary source of water.

Stripped of social, cultural, religious, and economic complexity, the ice stupa could be considered a simple ice reservoir. As such, critics argue

Figure 7.4 An ice stupa built during the competition of 2018–2019, with an ice café in the interior.

that it might be a better investment of time and energy on behalf of the environment to build massive artificial glaciers. The continued interest and application by NGOs and corporations in the artificial glacier projects in the region, and the more widespread use of this system by farming communities, is itself telling (Gladfelter 2018). The artificial glacier model, which relies upon community-driven buy-in and builds, avoids many of the limitations of the ice stupa while still harvesting significant amounts of irrigation water, as ice, for use the following spring.

However, as a design construct, the ice stupa is a straightforward structure that could be built by farmers, drawing from the water already allocated to each estate by the sophisticated and well-established rules for water governance in the region (Gutschow 1997). As such, the ice stupa design is a form of appropriate technology that enables farmers to control their own water resources, stashing a portion aside in times of abundance, for use in times of scarcity. This broad adoption was tested in the winter of 2018–2019 when the Ice Stupa Project ran a stupa building competition across Ladakh, and even more were built in the winter of 2019–2020 (Figure 7.5).

As a model, the Ice Stupa Project serves as a highly visible experiment in which authors have effectively drawn attention to the problem of climate change in high mountain subsistence agriculture communities, and then harnessed a complex array of benefits by building bridges across disparate

Figure 7.5 An ice stupa built during the competition of 2018–2019.

groups. In this way, the Phyang prototypes illustrate both the drawbacks of and the cause for complexity in climate-adaptive design thinking. The Ice Stupa Project highlights a host of noble goals, but also shows that multifaceted projects have more moving pieces than singular environmental interventions. This case study not only shows that intervening in an environmental system can trigger unintended consequences, but that if potential complications increase with additional layers of project complexity, then designers will need to envision, facilitate, and manage more moving parts in climate-adaptive design work.

Climate change design adaptations, such as the Ice Stupa Project, typically first become realized in an effort to improve environmental conditions, and then may also intertwine layers of social, cultural and religious meaning. The Ice Stupa Project is still working for acceptance in Ladakh, but the project also highlights many of the co-benefits possible through laminate framing, and the new directions that environmental activism may take, beyond ice reservoirs, in Ladakh. As this work continues to unfold in Ladakh, the international design community can learn from the creative partnerships that have been forged to organize the project and its rhetoric, and the limitations presented by competing stakeholder interests. In so doing, designers may forge new disciplinary space, by making the case for thoughtful, stakeholder-driven complexity in the broader collective project of climate change adaptation.

References

Angchok, Dorjey, Tsering Stobdan, and Shashi B Singh. 2014. "Community-Based Irrigation Water Management in Ladakh: A High Altitude Cold Arid Region." *Twelfth Biennial Conference of the International Association for the Study of Commons, Governing Shared Resources: Connecting Local Experience to Global Challenges*, 1–12.

"Artificial Glaciers of Ladakh | The Ice Stupa Project." n.d. Accessed December 29, 2019. http://icestupa.org/.

Darlington, Susan. 2012. *The Ordination of a Tree*. Albany: SUNY Press. www.sunypress.edu/p-5586-the-ordination-of-a-tree.aspx.

Devers, Quentin. 2017. "Charting Ancient Routes in Ladakh: An Archaeological Documentation." In *Interaction in the Himalayas and Central Asia: Processes of Transfer, Translation and Transformation in Art, Archaeology, Religion and Polity*, edited by Eva Allinger, Frantz Grenet, Christian Jahoda, Maria-Katharina Lang, and Anne Vergati. Vienna: Austrian Academy of Sciences Press. www.academia.edu/39101653/Charting_Ancient_Routes_in_Ladakh_An_Archaeological_Documentation.

Gagné, Karine. 2016. "Cultivating Ice over Time: On the Idea of Timeless Knowledge and Places in the Himalayas." *Anthropologica* 58 (2): 193–210. https://doi.org/10.3138/anth.582.T03.

Gagné, Karine. 2017. "Building a Mountain Fortress for India: Sympathy, Imagination and the Reconfiguration of Ladakh into a Border Area." *South Asia: Journal of South Asian Studies* 40 (2): 222–38. https://doi.org/10.1080/00856401.2017.1292599.

Gagné, Karine, Mattias Borg Rasmussen, and Ben Orlove. 2014. "Glaciers and Society: Attributions, Perceptions, and Valuations." *WIREs Climate Change* 5 (6): 793–808. https://doi.org/10.1002/wcc.315.

Gilmartin, David. 2003. "Water and Waste: Nature, Productivity and Colonialism in the Indus Basin." *Economic and Political Weekly* 38 (January): 5057–65. https://doi.org/10.2307/4414343.

Gladfelter, Sierra. 2018. "Ladakh's Artificial Glaciers, Ice Stupas, and Human-Made Ice Reserves." Fulbright-Nehru Student Research Report. United States-India Educational Foundation.

Gutschow, Kim. 1997. "Recent Research on Ladakh 6: Proceedings of the Sixth International Colloquium on Ladakh, Leh 1993." Edited by Henry Osmaston and Nawang Tsering.

"Ice Towers in the Desert – Rolex Awards." 2016. Rolex.Org. 2016. www.rolex.org/rolex-awards/environment/sonam-wangchuk.

Kolbert, Elizabeth. 2019. "The Ice Stupas." *The New Yorker*, May 20, 2019. Literature Resource Center.

Magnan, Alexandre. 2014. "Avoiding maladaptation to climate change: towards guiding principles." *S.A.P.I.EN.S. Surveys and Perspectives Integrating Environment and Society*, no. 7.1 (March). http://journals.openedition.org/sapiens/1680.

Maqbool, Raihana. 2019. "In Kashmir, An Ancient Solution Solves a New Problem." Global Press Journal. January 13, 2019. https://globalpressjournal.com/asia/indian-administered_kashmir/kashmir-ancient-solution-solves-modern-problem/.

Mingle, Jonathan. 2003. "Rewriting the Books in Ladakh." Cultural Survival Quarterly Magazine. December 2003. www.culturalsurvival.org/publications/cultural-survival-quarterly/rewriting-books-ladakh.

Nüsser, Marcus, and Ravi Baghel. 2016. "Local Knowledge and Global Concerns: Artificial Glaciers as a Focus of Environmental Knowledge and Development Interventions." In *Ethnic and Cultural Dimensions of Knowledge*, edited by Peter Meusburger, Tim Freytag, and Laura Suarsana, 8: 191–209. Cham: Springer International Publishing. https://doi.org/10.1007/978-3-319-21900-4_9.

Parvaiz, Athar. 2018. "Ice Stupas Spark Conflict between Farmers in Ladakh." The Third Pole. July 5, 2018. www.thethirdpole.net/en/2018/07/05/ice-stupas-spark-conflict-between-farmers-in-ladakh/.

Rizvi, Janet. 1998. *Ladakh: Crossroads of High Asia*. Delhi; New York: Oxford University Press.

Rogers, Everett. 2003. *Diffusion of Innovations*. 5th ed. New York: Free Press. www.amazon.com/Diffusion-Innovations-5th-Everett-Rogers/dp/0743222091/ref=asc_df_0743222091/?tag=hyprod-20&linkCode=df0&hvadid=312425492373&hvpos=1o1&hvnetw=g&hvrand=5993573091886049849&hvpone=&hvptwo=&hvqmt=&hvdev=c&hvdvcmdl=&hvlocint=&hvlocphy=9033030&hvtargid=pla-458892480208&psc=1.

Saxena, Siddharth. 2015. "Ice Stupas to End Water Woes." Times of India. February 24, 2015. https://timesofindia.indiatimes.com/india/Ice-stupas-to-end-water-woes/articleshow/46357875.cms.

Sharma, Arjun, and Kunal Bharat. 2017. "One's Waste, Another's Right: Translating History and Making the Ladakhi Commons." In *IACS Conference Proceedings*, 1–28. www.iasc2017.org/wp-content/uploads/2017/06/11N_Arjun-Sharma.pdf.

Strochlic, Nina. 2017. "The 'Ice Stupas' That Could Water the Himalaya." National Geographic. April 2017. www.nationalgeographic.com/magazine/2017/04/explore-desert-glaciers/.

Verma, Simant. 2018. "Meet the Team Making Man-Made Glaciers for Drinking Water." Red Bull. August 3, 2018. www.redbull.com/ca-en/theredbulletin/incredible-man-made-glaciers-ice-stupa-project.

Wangchuk, Rinchen Norbu. 2018. "This Ladakhi Ice Sculpture Unites Buddhism And Fight Against Global Warming." *The Better India* (blog). March 21, 2018. www.thebetterindia.com/135260/ladakh-ice-stupa-chorten-buddhism-global-warming/.

Wangchuk, Sonam. 2015. "Ice Stupa Artificial Glaciers of Ladakh." Indiegogo. 2015. www.indiegogo.com/projects/974995/fblk.

8 Snow barrier bands

Unlike artificial glaciers and ice stupas, snow barrier bands work by collecting water in the form of aggregated snow rather than ice. Masonry walls are placed at intervals along high passes to deflect snow into agricultural drainages. The term 'snow barrier band' was coined by the Leh Nutrition Project in 2013 when the NGO began to work on a major effort to resurrect the ancient practice of snow harvesting on Warila Pass. Others have called these systems snow fences (Dawa, Dana, and Namgyal 2000), snow block walls (Gladfelter 2018), or in Ladakhi, *kha gags*.

This form of snow diversion is an example of a traditional water management technique that happens to still be in use in Ladakh today. It is also an extraordinarily minimalist design solution, made up of materials gathered on-site, using low-tech engineering and no moving parts, and requiring very little lifetime oversight or management. As a design intervention that helps to mitigate climate change-induced drought, the snow barrier band stands out for its economy, utility and staying power (Figure 8.1).

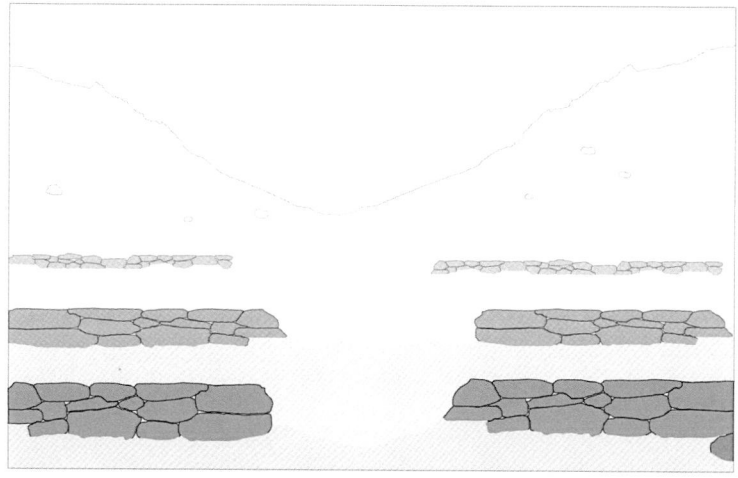

Figure 8.1 A diagram depicting a snow barrier band system.

Although the Ladakhi engineers at the Leh Nutrition Project call these projects "snow barrier bands," the design and application of this intervention is actually no different from a stone wall. It is the creative placement of the walls, meant to interact with the wind and the weather, that moves this design idea into the category of climate-adaptive design intervention. A ubiquitous feature in the village and agricultural landscapes of Ladakh, stone walls form animal and field enclosures, structural supports for houses and outbuildings, and in the form of the *mani* wall, even long, linear prayers. In this sense, the decision to use readily available stone material and accompanying local building knowledge to create a masonry snow break appears to be a natural, vernacular, fit for the region.

Background

The best example of a working snow barrier band in Ladakh today can be seen at Warila Pass, in the valley located just east of Leh. Here a series of masonry walls have been built to tip snowdrift from the Shyok River drainage into the more heavily populated Indus River drainage. The long, linear masonry walls line the upper saddle of the mountain's south-facing watershed, where they block snow from moving into the northern drainage on the other side, because winter winds typically blow from the south (Figure 8.2). Six different villages stand to benefit from this small-scale design intervention, as they harbor croplands well below Warila Pass, all of which rely on glacial and snowfield meltwater for irrigation.

Figure 8.2 A snow barrier band at Warila Pass.

The most recent reconstruction of this network, in 2013, reinforces the system of ancient walls that has been shored up at intervals over many centuries. Scholar Sierra Gladfelter's 2018 field research in Ladakh suggests that this particular system has long been a component of Ladakhi ice and snow harvesting efforts (Gladfelter 2018). She found evidence of much older walls at this site, as well as an ancient snow barrier band system on Changla Pass, above the village of Sakti. Gladfelter notes that although there is evidence of snow barrier band use over many generations in at least several places in Ladakh, the practice of using the collective labor system for their maintenance has all but disappeared in recent years. She suggests that when the first artificial glaciers were designed by Chewang Norphel in the late 1980s, and NGOs in the area began to funnel external funding and resources to irrigation projects, villagers in turn stopped managing systems for themselves. Involvement from the Watershed Development Programme, and area NGOs may have undermined village self-sufficiency, causing residents to give up traditional practices and involvement. Over the past four decades, outside programs have financed and organized much of the irrigation work in Ladakh, from the restoration of ancient systems to the deployment of new interventions.

While the state's Watershed Development Programme may have paid for much of the rebuild of the 2013 Warila Pass snow barrier band, stakeholder involvement still remains an important component of this design intervention. The Leh Nutrition Project organized a collaborative rebuilding of this ancient stone snow barrier band above the village of Sakti. At an elevation just over 17,500 feet, the Warila Pass snow barrier bands were built using stones from the site and some sweat equity from stakeholders. According to program facilitators, some 150–200 volunteers from six different villages camped and worked at the site, over a three-day period. While there were reports of villagers becoming sick from the altitude, requiring a number of people to descend, the completed wall was successfully rebuilt. The wall they reconstructed had been dormant for approximately 30 years, and when finished, resurrected a much older system. The project is now complete, and it features at least nine different walls that range in size from 40 feet to 100 feet in length.

How it works

The snow barrier band is a system of freestanding masonry walls used to contain and redirect snow. These walls are built on high mountain passes where the prevailing winter winds would otherwise blow snow drifts across into a northern drainage, causing the windswept snow to move into the (usually uninhabited) drainage below. Instead, the snow barrier band blocks that snow from moving to the north side of the mountain, causing it to pile up on the south side of the drainage and eventually melt into the south-facing watershed below (Figure 8.3). When placed strategically

Figure 8.3 The wall, made up of piled rocks, effectively blocks windswept snow.

across a landscape, these low masonry walls can effectively redirect snow-fields from one watershed to another (Figure 8.4).

Snow barrier bands perform much like the snow fences found in other parts of the world. Snow fences are prevalent in other countries to stop snow from blowing horizontally. These fences often abut roads on wind-swept plains, where the movement of snow onto a road would limit mobility. Snow fences are used to stop and hold snow, managing the buildup of snow to maintain a clear space on the leeward side of the fence. It is unusual, however, for these interventions to also use the snow that aggregates on the windward side of the fence. Unlike snow fences, where snow is treated as a problem, snow barrier bands view the accumulation of snow as a benefit, worthy of collection.

In Ladakh, snow barrier bands is composed of long, linear masonry walls built at high mountain passes in the upper portion of a watershed. The walls are typically built from stone found on the site, standing three to five feet high, one to two feet deep, and many hundreds of feet in length. Individual walls act as a part of a larger series, where in aggregation they help to capture snow, hold it in place at the top of the watershed, and then release it gradually in the form of meltwater in the late spring.

Most of the material used to produce a snow barrier band is found on site, in the form of rocks. This allows the walls to be very affordable, but limits their placement to areas where rock is available. The primary cost of

Figure 8.4 Fields below Warila Pass benefit from additional meltwater produced by snow barrier bands above.

snow barrier bands now comes from labor, if it is hired out, rather than volunteered by villagers. Occasionally, reinforcing wire and concrete will be used to augment a system, which increases its overall cost.

The stone walls that make up a snow barrier band improve village water security by collecting and conserving snow that would otherwise flow away from farms as meltwater. The additional snowmelt provided by snow barrier bands increases the overall water allocation available to farmlands located directly below the system. Mounded snow reserves that have been trapped by snow barrier bands begin to melt early in the spring, giving farmers a known quantity of water that can help jumpstart the planting season. This is particularly useful in the context of Ladakh, where high-altitude farmers must wait for spring snowmelt before sowing their fields, and in places where village communities are dependent upon minimal winter snowfall for the bulk of their crop irrigation. The meltwater produced by snow barrier bands may effectively give farm irrigation capacity a boost, enabling more planting overall, or spurring earlier farming activity, both of which could in turn increase agricultural production (Dawa, Dana, and Namgyal 2000).

According to engineer Chewang Norphel, an individual five-foot-high wall can create roughly 15 feet of snow buildup on the south side of a pass, most of which would otherwise blow into an unpopulated northern watershed. Without specific field data, however, this anecdotal report still needs to be verified. According to Dawa et al., incremental improvements to the ancient wall systems at Warila Pass in the 1980s created a clear irrigation benefits in the minds of village stakeholders, although again the actual capacity increase was unmeasured (Dawa, Dana, and Namgyal 2000, 244).

Scale

Snow barrier bands operate at the scale of an individual wall and extend in scope and range to the watershed and beyond. These systems reveal design thinking that demands the consideration of an entire valley and the careful understanding of environmental factors, including wind direction, solar aspect, snowfall patterns, and the physical locations of downstream villages.

In Ladakh, many villages have been built on the south side of mountain ranges, in order to take advantage of passive solar heating potential during cold winter months. In many cases, the winter winds blow from south to north, effectively propelling valuable winter snowfall from the densely populated side of the mountain range to the uninhabited northern slope. Snow barrier bands serve to stop snow before it can tip from a south-facing drainage into a north-facing drainage, thereby increasing the amount of meltwater that will be available to farmers in populated areas. This scheme works in part because of the overarching demographic trends of the region, with more populous southern slopes.

In this way, the snow barrier band may improve water security at the scale of the village, but not necessarily at the scale of the mountain. The reapportioning of snow, in this case redirecting snowfall to feed one area rather than another, will not increase the overall amount of snow in the region. Indeed, a boost to one watershed will necessarily deprive another watershed. The snow barrier band has historically worked with success in Ladakh, primarily because the region is so sparsely populated that the gains of one drainage do not necessarily appear to be significant losses in another drainage. As a design idea, the snow barrier band is thus limited in terms of deployment and overall capacity, and could also have limited application in other areas.

Benefits

Snow barrier bands are scale-able: they can be made up of just a few walls to corral snow into an isolated space, or they can work in aggregation, stretching out across enormous landscapes for broader irrigation management. This flexibility with regard to size enables a certain economy, as well as the possibility of real affordability and incremental production. Moreover, a section of barrier band walls demands a small investment of construction material and time, relative to artificial glaciers and ice stupas.

Snow barrier bands are also derivatives of ancient local building vocabularies. They incorporate masonry walls, which constitute the vast majority of traditional Ladakhi construction, from houses and temples to animal enclosure, crop protection, and fences. This is not only a construction type that is understood and easily built by local people, but it is also a visual fit with other structures in Ladakh's high mountain context.

The location of snow barrier bands, atop high mountain passes, also provides an opportunity for collaborative stewardship of the watershed. While NGOs have in recent years taken over the management and organization of wall work parties, snow barrier bands have always brought people together. Because they serve an entire watershed, whole villages will work together to construct and maintain these systems, and they require the buy-in of many different stakeholders. This type of collective interest and engagement can serve as a catalyst for other types of projects among stakeholder groups.

Limitations

The sheer scale and impressive altitude of a snow barrier band project may also be its largest limiting factor. Work parties must collect stones and move this rock into place at a site that can be 17,500 feet. This altitude, combined with the considerable length or size of a typical snow barrier band, presents a significant challenge for new construction. However, it is worth noting that roads in Ladakh are built under the same difficult conditions, demonstrating that this work can be done. Moreover, once a road crosses a high-altitude pass, earth-moving machinery could potentially access snow barrier band sites to assist in construction in the same location.

Because this project relies on age-old methods for engineering, construction, and deployment, there is little risk of negatively impacting other systems. Still, there is the potential to cause negative downstream effects, particularly as this type of work scales up across a watershed. For instance, snow barrier bands have been built at high mountain passes where diverted snow drifts would otherwise fall into uninhabited drainages. The collection of snow and diversion of meltwater into populated valleys is something that has been in place for many decades, if not centuries. Because of this long tenure, snow barrier bands do not appear to drastically alter the experience of other people. If new snow barrier bands were placed at mountain passes where both sides depended on snowmelt for agricultural functioning, then the relative difference in irrigation access caused by the placement of a new system might be more visible, and also problematic.

The development of additional snow barrier bands will need to be done with full awareness of the impact on other systems, places and people. Indeed, each time a new design intervenes in a system to divert water, designers must consider what purpose that water held before, and how the lack of water will be felt downstream. The walls of the snow barrier bands appear to enable movement of animals, but the water that is redirected could impact any number of animals and plants found in a system's northern drainage (Figure 8.5). A chief concern of any new snow barrier band

Figure 8.5 Yaks cross Warila Pass unencumbered by the position of numerous snow barrier bands.

project must necessarily be potential damage to other human or natural systems.

In Ladakh, one population group with a very limited voice in such decisions are the nomadic herders, landless tribal people who have a right to use the mountain hinterlands for livestock grazing and their own itinerant camps (Prakash 1991). These people may, for instance, be actively using "unpopulated" drainages without actually having a strong visible presence in that space. Here the nomadic herders could potentially notice a decline in water availability caused by the deployment of new snow barrier bands diverting snowdrifts from high mountain passes. Moreover, marginalized groups may not have the voice, platform, or right to effectively lobby for their needs, particularly in comparison to those of village land holders.

Moving forward

Unlike the slick engineering built into artificial glaciers and ice stupas, snow barrier bands are simple architectural constructs; they perform more like the snow fences used in North America than the complicated water harvesting schemes elsewhere in Ladakh. While the snow barrier bands employed for water redirection in this region have been made of stone, they could in theory be constructed from any number of materials: living plants, imported plastic webbing or metal screens, hay bales or sticks. The materials used in Ladakh reflect the needs and opportunities of this landscape: an abundance of site-strewn stones, the high cost of imported goods, and the hungry herds of animals that roam these spaces. Although snow barrier bands typically consist of an upright wall, they could also be dug into the landscape to a lesser effect, forming a "snow pit" with the same aims and techniques.

Just as the scale of the snow barrier band extends across a mountain watershed, so too must the decisions involving the deployment of new systems. Ladakhi village stakeholders often make decisions about climate-adaptive design interventions using a consensus decision-making model. In many of these decision-making processes all of the village stakeholders are invited to participate, with household representation that includes men and women, elders, and young farmers. During the course of Ladakhi history, inter-village decision-making has been overseen by elected officials, colonial powers, village headmen, and monastic leadership. More recently, these discussions have been facilitated by NGOs such as the Leh Nutrition Project.

While the most recent work involving snow barrier bands in Ladakh happens through NGO programming, and indeed exhibit many of inclusive, accessible, and egalitarian decision-making characteristics, not all stakeholders have been invited to participate. The nomadic herders of Ladakh may well provide valuable insight into future planning projects, and also arguably deserve to have veto power for projects that negatively impact their own livelihoods. To a similar extent, other stakeholder representatives, such as organizations protecting animal and environmental interests, might be invited to participate. If new snow barrier band projects are proposed in Ladakh, it will be increasingly important to create an open discussion and decision-making process that includes all types of interest groups.

Conclusion

While only two snow barrier bands have been documented in Ladakh by researchers, others may exist, whether in Leh District, or the nearby mountains of Chinese-controlled Tibet and Pakistan. Regardless of these unknowns, these design responses to the region's chronic drought are still considered relatively rare. Today, the design and construction of new snow barrier bands largely requires an infusion of government or NGO support. However, because they are constructed from regionally ubiquitous materials using local labor and construction methods, these design interventions in Ladakh present a relatively affordable alternative to other types of proposals, including concrete reservoirs, irrigation canals, and pumping systems. As a response to climate change induced water scarcity, and especially in the cold desert agricultural context of Ladakh, snow barrier bands could scale up to provide many other villages with an additional source of water for crop irrigation.

In the context of the remote Indian landscapes of Ladakh, agricultural officers and hydraulic engineers have not definitively measured the amount of water annually coursing through each of these systems, much less their ability to improve irrigation functioning. Indeed, accounts of efficacy have been mostly anecdotal, and the growing body of relevant publications remains largely limited to reports from visitors. Admittedly, even the age-old practice of redirecting snow needs to be understood as an unproven design idea, and parameters for success and failure still need to be established.

More fundamentally, while snow barrier bands may make good use of existing natural water resources, they cannot conjure up new reserves. The decoupling of precipitation from seasonal norms under climate change has shifted the amount of water falling as snow annually in Ladakh (Mingle 2015; Nüsser et al. 2019). Increasingly, precipitation comes to the area as rain in the summer months, occasionally in cloud bursts that drop more water at one time than can be used or stored. In this context, the development of snow barrier bands may stretch a shrinking resource by harvesting useful winter snowfall, but does little to harness abundant summer precipitation.

However, proponents of these traditional climate-adaptive design interventions could argue that they address water scarcity up to a certain point, and as long as they mitigate the effects of climate change locally, they provide a much-needed interim solution for struggling farmers. Snow barrier bands reflect an effort, by farmers arguably standing on the front lines of climate change, to extend and control their limited water reserves.

The snow barrier bands of Ladakh represent a climate-adaptive design intervention that marries age-old irrigation practices with new notions of civic agency and environmental activism. The use of gravity, siting, and stone walls places the work squarely within traditional agricultural practices in the region. However, new partnerships with NGOs such as the Leh Nutrition Project and the use of workshops and work parties signal a fusion of old methods with more contemporary participation techniques. If this work is to continue, an even broader array of stakeholders will need to join in the effort, and because the funding for snow barrier bands has become absorbed by state and non-profit groups, this vision may actual be possible.

References

Dawa, S., D. Dana, and P. Namgyal. 2000. "Water Harvesting Technologies and Management System in a Micro-Watershed in Ladakh, India." In *Waters of Life- Perspectives of Water Harvesting in the Hindu Kush-Himalayas: Volume II*, edited by S.R. Chalise and M. Banskota, 235–59. Kathmandu: ICIMOD. http://lib.icimod.org/record/22407.

Gladfelter, Sierra. 2018. "Ladakh's Artificial Glaciers, Ice Stupas, and Human-Made Ice Reserves." Fulbright-Nehru Student Research Report. New Delhi: United States-India Educational Foundation.

Mingle, Jonathan. 2015. *Fire and Ice: Soot, Solidarity, and Survival on the Roof of the World*. 1st edition. New York: St. Martin's Press.

Nüsser, Marcus, Juliane Dame, Benjamin Kraus, Ravi Baghel, and Susanne Schmidt. 2019. "Socio-Hydrology of 'Artificial Glaciers' in Ladakh, India: Assessing Adaptive Strategies in a Changing Cryosphere." *Regional Environmental Change* 19 (5): 1327–37. https://doi.org/10.1007/s10113-018-1372-0.

Prakash, Sanjay, ed. 1991. *Solar Architecture and Earth Construction in the Northwest Himalaya*. Sustainable Development Series 5. New Delhi: Har-Anand Publications in association with Vikas Pub. House.

9   Solar design

One of Ladakh's clearest environmental assets comes in the form of solar radiation; the region receives more than 300 days of full sun per year on average, annually exceeding other areas in India (Prakash 1991). This translates to nearly 7000 Whr/m²/day on the horizontal plane: it is a significant source of reliable energy (Wanchuk, 2019). In the context of Ladakh's cold desert landscape, access to a free, clean, readily available heating and power resource is not to be underestimated.

The use of the sun as an energy and heating source is well-documented in the long history of Ladakhi architecture, although until recently this use was largely limited to passive solar radiation (Prakash 1991). Starting in the early 1980s, a number of NGOs active in the area began working on various projects that improved efficiency in lower-tech solar solutions.[1] In addition to these "improved low-tech" solar investments, new types of technologies, such as village-scale photovoltaic arrays and photovoltaic light fixtures, have been introduced to the area. These three approaches to solar design in Ladakh ("low-tech," "improved low-tech," and "high-tech") each play a role in the region's response to the pressures of climate change.

The use of solar heating and power is not necessarily tied to climate change adaptation. However, as villages seek out energy sources that will not further degrade the planet, and as residents work to create alternative lifestyles that might be sustained under resource scarcity caused by climate change, solar design interventions play an important adaptive role. After all, the accepted cause of climate change (which stems from greenhouse gas emissions) could be mitigated in part by moving away from burning dirty fuel for heat and power. In Ladakh, moreover, it is becoming increasingly apparent that sustainable and self-sufficient livelihoods under climate change projections may actually hinge upon the use of solar design interventions.

Background

Ladakhi solar design techniques operate in both active and passive ways. Active solar arrays provide electricity that is primarily used for lighting, heating, pumps and other electronics. These electricity arrays can be individually mounted to building rooftops, or aggregated in one location with connections to multiple users. In Ladakh these costly systems tend to be either wholly or substantially subsidized by the Indian government, with associated outsourcing of the design, construction, and ongoing maintenance. Ladakhi communities benefit from these "high-tech" solar strategies in terms of improved access to clean, affordable power, but individual stakeholders tend to lack agency in terms of broader decision-making and system accountability.

Passive solar systems engage architecture more directly, incorporating a building's shape, orientation, and materials to harness (or deflect) sunlight for heat (or coolth). Many Ladakhi villages have been strategically placed in south-facing drainages to increase their access to solar gain across an entire community.[2] At the scale of the building, many households orient to the sun, and then organize specific rooms within these structures to benefit from relative warmth or coolth. Finally, materials designed to trap or avert solar radiation have been used in the makeup of individual buildings, comprising cladding, roofing, or wall assemblies. Together, these techniques represent the scale shifts and variety of outcomes that are available in Ladakhi passive solar projects.

Currently most Ladakhi households are poorly insulated, which leads to uncomfortably cold ambient interior temperatures, high heating costs for imported fuel, or a reliance on large quantities of hand-harvested fuels for burning (Mingle 2015). Collected dung has served, in many areas, as a primary source for heating and cooking fuel; there are both human and environmental costs to using this harvested fuel source. In the former case, poorly engineered stovepipes and the use of a dirty burning fuel cause serious health hazards for household inhabitants (Mingle 2015), and in the latter, dung burning releases greenhouse gas emissions and air pollutants.

Alternatives to dung burning can be found in many villages, provided by imported propane tanks for heaters and cook stoves, for example, diesel brought in to run generators, and kerosene fuel used in lamps. Some large villages or towns, such as Leh, and neighboring Choglamsar, also run larger power grids using a combination of hydropower and diesel generators. As with dung burning, the ubiquitous generators increase greenhouse gas emissions, and they also trigger other complications such as noxious fumes and unpleasant ambient noise. Unlike dung burning, most of these alternative fuel options are relatively new additions to Ladakhi villages, having only been introduced in the 20th century, and in many ways they offer clear benefits to households, particularly in terms of convenience and ease. But non-harvested fuels must also be brought in to remote villages over many miles of rough road, and they can be costly expenditures for Ladakhi households.[3]

There are many reasons for seeking out sustainable heating and electrical options for Ladakhi communities. The burning of imported kerosene,

Figure 9.1 Dung dries on a rock outside a village in Ladakh, to be used for fuel later.

propane or diesel, and locally harvested animal dung degrades the region's immediate air quality and contributes more broadly to climate change. Imported fuels have numerous drawbacks, including access, cost, and noise. In terms of logistics, winter weather can cause road closures for months at a time, effectively halting fuel deliveries. In terms of equity, it is worth considering that a number of villages operate outside of a cash economy and many individual households simply don't have the resources to pay for imported power. Meanwhile, traditional fuel sources also come with significant disadvantages. Dung collection is a laborious process made more difficult by the dwindling numbers of livestock under climate change adaptation. According to a 2004 report by Vincent Stauffer, Ladakhi "Women spend two months a year on average collecting dung from the pastures for cooking and heating" (Stauffer 2004, 3) (Figures 9.1 and 9.2). Worse, the practice of dung burning is now recognized as unsafe; women and children in particular suffer from respiratory problems caused by incomplete incineration on interior cook stoves and fireplaces (Mingle 2015). In pure economic terms relating to sustainability, the dominant electrical and heating methods currently practiced in Ladakh cannot continue at current rates. As resource scarcity under climate change reduces the numbers of animals that can be sustained by the landscape (contributing to an overall reduction in dung availability), and as fuel reserves become more scarce and expensive, alternative energy sources are clearly needed.

Figure 9.2 Collected dung, brought in from fields and dried, constitutes a major fuel source for villages in Ladakh.

"Low-tech" solar responses

Traditionally, Ladakhi houses and settlements were designed to reflect the region's environmental constraints through their physical forms and functional roles. Just as the limited supply of water available in the region has led to sophisticated irrigation management strategies, the lack of abundant fuel has left an imprint on vernacular form by directing the shape of Ladakhi buildings and villages (Norberg-Hodge 1991). Long, cold winters and a relative scarcity of heating fuel have necessitated the compact form of Ladakh's masonry structures, many of which feature both multi-story living and cohabitation arrangements for livestock (Figure 9.3). Most historic buildings, and indeed entire settlements, have been designed to take advantage of the sun through passive solar orientation, by orienting fenestration to the south whenever possible (Ferrari 2018). Finally, the region's tenuous road connections and the high cost of transportation is reflected in the widespread use of locally-sourced materials and construction techniques. In using building mass to store heat, in orienting to the south, and in incorporating local building materials, the Ladakhi vernacular exhibits "low-tech" solar design.

As a design approach, passive solar orientation has been codified by Ladakhi social and cultural norms, as well as through dominant construction techniques. Historically, traditional Ladakhi buildings responded to

Figure 9.3 Vernacular architecture in Ladakh.

basic passive solar rules of thumb, primarily by opening up fenestration to the south side, and placing areas of active use along the southern side of the building. Areas of infrequent use, or rooms where residents spent a minimal amount of time, have historically been placed in the interior or on the northern side of the building. Spaces such as store rooms and toilets tended to be located away from direct sunlight, and these areas often lacked windows as well.

A number of rooms serve specific functions in the Ladakhi household, and these spaces frequently also have a solar relationship. These areas play an important role in social gatherings and religious functions, referencing broader cultural values. While most of these spaces can be referenced in the earliest documentation of Ladakhi structures from historical accounts, photographs and field observation, subtle shifts to these standards also occurred over time. For instance, early vernacular structures used paper and wood lattice windows for light, as glass was not widely accessible (Ferrari 2018). Once glass window panes were brought into Ladakhi villages in the second half of the 20th century, building facades shifted to incorporate this new material, which then helped to highlight the types of rooms inside.

Today, as in the past, the rooms most closely associated with solar heat and sunlight are those reserved for frequent use and for guests. The *shelkang* (solarium) is a heavily glassed, often south-facing room typically reserved for visitors and special occasions. A *donkhang* (guest room), may also have south or west facing windows. Typically, the *chansa* is a south-facing living and dining room where a Ladakhi family may spend long hours indoors together during the cold winter months (Ferrari 2018).

While many of these south-facing rooms are ubiquitous throughout Ladakh and serve as an indicator of wealth or household means, they also typically lack insulation for effective heat capture. In the summer, daytime use of these spaces can be unbearable due to the heat trapped by glazing, while in the evenings (especially in winter months) these same spaces will dip below freezing. While solar design is a helpful vehicle for the thermal management of these areas, effective solar heating outcomes require additional tools for adjusting these spaces daily and seasonally, as well as insulation and a tight building envelope, particularly because Ladakh is located in a geographic context with strong diurnal swings (Prakash 1991).

Even in vernacular Ladakhi buildings, it is not uncommon to see thermal comfort surrendered for the sake of propriety or for aesthetic values (Ferrari 2018). The general technique of opening building facades along the south side is an example of how many villages and households work with the sun to maximize winter warming while reducing household fuel consumption. But the challenge of maintaining thermal comfort through swings in daily and seasonal temperatures cannot be met through glazing alone. Instead, additional techniques for insulation and shading may be overlaid onto passive solar orientation strategies for improved outcomes.

"Improved low-tech" solar responses

A host of "improved low-tech" strategies for reinforcing or bolstering building performance have been developed by enterprising Ladakhis with the advent of new materials in the region, and also brought in by NGOs, visiting experts, and through government programs (Prakash 1991). These ideas have typically been tailored to the region, often acknowledging the limitations of the area with regard to cost, construction and transportation, as well as opportunities afforded by both climate and context. While many different ideas have been explored in the past five decades, the design interventions featured in this section are those that have exhibited the most popular interest or staying power.

Insulation

This basic construction technique is, somewhat incredibly, not well represented in traditional construction practice in Ladakh. This is a surprising outcome because the use of insulation is an absolutely critical component of thermal control with passive solar gains, especially in cold climates. In recent years NGOs and visiting experts have worked with village stakeholders to test various types of insulation techniques, usually by placing lightweight materials in wall cavities or sandwiching them into roof assemblies. Materials used for insulation range from local agricultural and animal bi-products such as wool or straw, to imported materials such as spun fiberglass. Similarly, precision machined windows, doors, and ventilation devices have been brought into the region to produce tight building envelopes. These two strategies – incorporating insulation and tightening the building envelope – help to reduce heat loss and improve user control, allowing passive solar strategies to work more effectively.

Shading devices

Traditional Ladakhi vernacular buildings tend to have flattened facades without major projections. However, shading devices, particularly on the exterior of east, south and west facades, can help to mediate solar gains. A number of newer building projects in Ladakh incorporate shading devices that block harsh summer sunlight without impacting winter warmth. Because these exterior devices rely on imported materials and technologies (such as prefabricated awnings, or metal fasteners to tie protruding sections into the building structure) they tend to be used in public projects by visiting designers, NGOs, or the government.

Trombe wall

The trombe wall was first introduced in Leh in 1979, by the London-based Intermediate Technology Development Group (Tonge 1982). Working in conjunction with windows, a trombe wall can be built alongside any a large, south-facing glazed wall. The trombe wall is a design response that is meant to address the diurnal temperature swing, allowing solar gain during the day to be stored in a massive wall, which then can radiate heat back throughout the night. To achieve this, a glazed south-facing opening must be paired with an interior wall of considerable mass. In Ladakh, these interior (trombe) walls stand about one foot inside the exterior wall of windows, made of masonry and painted black to absorb heat (Figure 9.4). The trombe wall usually is placed directly in front of an array of southern windows, with small vents placed throughout the wall to enable daily management of the flow of heat. In placing the walls in front of the windows on the south side, views and light are blocked to the interior.

NGO aid groups have been extremely supportive of the trombe system, sponsoring building workshops, paying for retrofits, and creating models in public buildings throughout the region (Hales 1986). However, while these walls exist in households and businesses throughout Ladakh, and certainly do work, they haven't been as popular as one would hope or

Figure 9.4 The trombe wall at SECMOL faces south to capture sun during the day.

imagine. Ladakhi homeowners dislike the large black walls, and cite the associated lack of light and views as one major drawback (India Today 1994). While these massive features do indeed block views to the outside, they still perform well, as large passive heating sources that can considerably alter the thermal comfort of the interior room.

A number of NGO-sponsored public buildings, such as health clinics, schools, and libraries, have incorporated the trombe wall heating strategy to good effect (Figures 9.5 and 9.6). Unfortunately, these types of spaces rarely see use during the evening hours, so the system's significant thermal benefit (in the form of evening warmth) is ultimately under-experienced. In household applications, where the trombe wall makes the most sense due to patterns of use, most Ladakhis have preferred to retain the large glazed walls of specialty rooms for aesthetic qualities (Joshi 2013). While the trombe wall continues to be supported by NGOs and some individuals, the loss of views and the strange look of a large black masonry wall in interior rooms appears to be too great a barrier to widespread adoption.

The trombe wall technique could, however, be implemented in other types of rooms where warmth is valued over light and views. Indoor first-story livestock areas, for instance, or the massive masonry composting

Figure 9.5 Inside of a patient room, at the clinic at Sani, a trombe wall holds heat during the day.

toilet blocks both usually have a south-facing wall and could easily incorporate this strategy. As Ladakhi households change over time, creating new walls, rooms or even new buildings, the trombe wall technique may find new opportunities for application and better regard among stakeholders.

Figure 9.6 The clinic at Sani, under construction, to use trombe walls facing south.

Hot water

Cooking with solar power is a relatively new "improved low-tech" design idea in Ladakh, brought in by visiting designers in the 1990s, and popularized in the intervening decades. In this technique, small mirrors line a moveable concave surface to create a sun tracker that can focus the sun's rays on a single point. The mirrors reflect the sun, pointing it toward a perch where a pot of water, or a kettle may be placed. This technique is used all over the world, in tropical as well as dry environments, as a means of providing households with passive water heating. However, in Ladakh this technology eliminates the need to heat water on a stove using any type of fuel, and while it typically takes more time to produce hot water than other conventional means, this work can be done without oversight, outside of the house, during the course of the day. Moreover, because the intensity of the sun is so strong in Ladakh, water can heat relatively quickly when made in this way.

The mirrored water heating devices have been brought into Ladakh by a variety of aid groups, and it is not uncommon to see them placed outside of houses and businesses, even far from roads in remote villages. Large schools, such as SECMOL (Students' Educational and Cultural Movement of Ladakh) in Phey, host enormous solar cooker arrays, standing up to 10′ in diameter (Figure 9.7). However, a typical household array tends to be sized with a 4–5′ circumference, and this is adequate for a single family's daily cooking needs.

Figure 9.7 A solar hot water cooker, made up of mirrors, is one of two large arrays
at SECMOL.

Larger stores of hot water have only recently been needed in Ladakh. As
tourists began to visit the area over the course of the last five decades, guest-
houses and hotels responded to the perceived needs of visitors by offering
hot showers. The present-day popularity of Western-style showers then led
to the development of a new "improved low-tech" design, the solar hot
water heater (Figure 9.8). These arrays are typically placed on rooftops, and
in this way the adaptation strategy has become increasingly visible to the
outside world, especially in Leh and other areas where tourists might enjoy
hot showers. Most of these heaters are commercially produced outside of
Ladakh, which impacts access in terms of cost, maintenance and access.
However, in many villages and towns, especially where large amounts of
hot water are needed to support tourists, solar hot water heaters have provided a relatively clean-energy, affordable design solution.

"High-tech" solar responses

Lighting

Solar lighting programs offered by the state do much to alleviate the electricity demands of off-grid villages in Ladakh. A household lighting array

Figure 9.8 A solar hot water device, at SECMOL, produces hot water for showers.

may consist of just one or two fluorescent light fixtures, attached to the ceilings in well-used rooms, and connected via wiring to a small rooftop photovoltaic array. Small, integrated batteries enable evening lighting long after the sun has set. These household-scale interventions can be dispersed and installed at a relatively low cost to the Indian government, and they satisfy some of the region's most pressing electrical needs by illuminating Ladakhi households.

Village-scale photovoltaic arrays

Within India, Ladakh is recognized as a strategic location for active solar harvesting. Its location "on the roof of the world" provides extraordinary solar access. The region's two contested borders underscore the need for village resources that support national security. Finally, the rugged and mountainous terrain makes it difficult to build and maintain long utility connections, calling for a more discrete and self-sufficient system. More than 85 percent of Ladakh is served by limited photovoltaic electricity, which has come to be seen as an affordable and accessible government-funded means of electrifying the region (Jacobsen 2000).

For years, the Indian government has invested in small-scale clean power solutions for Ladakh through small photovoltaic power arrays. These arrays act as micro-energy plants in remote villages. Unlike the imported fuel used to run stoves and generators in these communities, photovoltaic electricity is clean, quiet, and produced on site. These new systems are preferred by most stakeholders, as they reduce the cost and work burden on households. But because the plants must be brought in, built, maintained and paid for by the government, they remain out of reach for some communities. Currently villages in Ladakh may request solar power assistance from their local political representatives, but they have no real control over who will receive those benefits from year to year. Once a solar electricity array is built in a village, it is typically sited so that the majority of village households can be served.

In 2019, the Indian government announced that it plans to build the world's largest power array in the mountains of Ladakh (Dutta 2019). Ladakhis stand to benefit from this massive power array, which will reduce dependence on dirty burning stoves and generators. This investment promises to lead to healthier air quality for villages served, as well as an efficient, affordable and quiet on-demand electricity source. With the addition of affordable electricity, photovoltaic power plants provide additional benefits to the economy, safety, and possibly even agricultural productivity. This scheme for power production stands out as one of the most significant transitions to clean power in Ladakh to date, and as such, provides an example of government-led climate-adaptive design assistance.

Materials, technologies and trade

Historically, the Ladakhi vernacular form was born out of a combination of material availability, climactic needs, and social or cultural preferences. Without access to a source of abundant and various trees in the region, ceiling spans, column size and window frames reflected the physical

Figure 9.9 At Sani clinic, battery modules for PV electricity have been made possible by NGO aid.

limitations of locally harvested Willow and Poplar trees. Because steel, glass, and concrete were also difficult to source, vernacular buildings primarily incorporate local rock and mud, and construction techniques tend to preference hand construction. Material availability continues to be a driving factor in architectural design in the region, where specific formal features derive from an extremely limited palette. With regard to solar design possibilities, widespread adoption hinges upon access to materials and technologies, the cost of new systems, and available subsidies.

Climate-adaptive design projects featuring solar heating or electricity are not uncommon in Ladakh, and indeed even the most basic new buildings tend to reference and benefit from a relationship to the sun. However, more advanced techniques for solar use in construction, such as the use of technologies associated with photovoltaic power arrays, and tools used in building commissioning, remain out of reach of most village stakeholders. The high cost of imported materials and technologies, combined with a dearth of associated repair sites, will limit widespread use. However, government and NGO initiatives have brought some of these costly techniques within reach of Ladakhi villagers (Figure 9.9). These alliances may actually prove to be co-beneficial, as the government can avoid bringing electricity to towns via a utility grid or with generators in places where solar already manages the need.

Financing and alliances

Many institutions, working with funding and expertise from the Ladakh Renewable Energy Agency, have had success with passive solar design strategies. These sites include hospitals and healthcare facilities, government

Figure 9.10 The Druk White Lotus educational campus, designed and built in partnership with ARUP, uses passive and active solar in many buildings, including at the composting toilet block pictured here.

Figure 9.11 The Druk White Lotus educational campus, designed and built in partnership with ARUP, uses passive and active solar in many buildings, including at the classroom pictured here.

buildings, schools and community buildings. Other projects, funded exclusively by local NGOs, have also made extensive use of passive and active solar harvesting (Figures 9.10 and 9.11). These projects point to a design trend whereby the construction of passive solar community buildings begin to reflect a new aesthetic and environmental attitude for Ladakh.

In an effort to promote solar techniques, the Ladakh Renewable Energy Agency oversees the development of many of the solar projects in Ladakh, as a branch of the Indian government's Ministry of New and Renewable Energy. They consult with Indian corporations to execute the work, and subsidize between 50 and 100 percent of the cost of the work (Daultrey and Gergan 2011). In addition, at least six different NGOs in Ladakh collaborate on this topic through a coalition. Together these groups offer training through community meetings, workshops and model houses. One example of this coordinated effort to provide solar development across Ladakh can be seen at the Students' Educational and Cultural Movement of Ladakh (SECMOL). Contracted by Groupe Energies Renouvelables, Environnement et Solidarités (GERES), SECMOL has trained rural engineers in the methods of solar building for several decades. Finally, SECMOL stands out as a model campus for demonstrating the working value of such systems: the school uses only passive solar on its residential campus, which is notable when considering that the program has offered winter residential programs for more than 15 years.

Moving forward

Entirely new technologies, such as the hybrid solar-wind systems that have been piloted in Ladakh, point to the region's interest in adopting new ideas. At the same time, considerable effort is being invested to improve and modify existing systems that incorporate solar design thinking. In this way, the solar design interventions represent a fairly wide-ranging array of responses, all specifically suited to their context in Ladakh. For instance, the Ladakh Renewable Energy Development Agency (LREDA) has worked closely with villages to survey their needs before initiating new systems or technologies (Daultrey and Gergan 2011). Additional stakeholder involvement, especially in determining the right fit for future projects, is needed (Heath 2015). Solar heating initiatives, for instance, are still underfunded in comparison to solar lighting projects in the area.

As the road network in Ladakh is expanded, new technologies and products will be able to more easily reach village consumers. At the same time, the high cost of new technologies and ideas is diminishing, as products become more efficient and mainstream. The lower cost of going solar, as well as the support provided by government and NGO initiatives, will

Figure 9.12 New construction, using vernacular forms, opens windows to the south side for solar benefit.

Figure 9.13 New construction with contemporary materials and form uses solar orientation.

likely Ladakh's improve access to development benefitting from solar orientation in the coming years (Figures 9.12 and 9.13).

Conclusion

Solar use in Ladakhi villages will likely continue under resource scarcity, and may even help to mitigate climate change. However, relatively few new buildings currently prioritize solar benefits to the exclusion of aesthetic values, even though the technology, awareness, education, certified builders and environmental conditions make a case for solar design. Subsidized energy reserves through the Indian government, a reliance upon (and belief in) newer materials such as concrete and steel, and a general move toward Western-style buildings has all interfered with widespread solar adoption.

However, if climate change continues to impact energy reserves in the region, one would expect to see the adoption of an even more environmentally-responsive architecture. Building orientation to the south, with windows along that side and a long east-west axis, is not only already the norm, but also readily achievable for most new buildings. To improve the functioning of passive solar design, this low-tech solution also needs moveable components for regular façade management, building insulation and a tight envelope. In conjunction with this low-tech design thinking, the overlay of other "improved low-tech" instruments to assist in solar gain (such as trombe walls, solar cooking devices, and solar hot water heaters) can quickly improve building energy use. Finally, "high-tech" solutions such as photovoltaics may launch Ladakh into net neutral territory. In so doing, these alternative approaches suggest the creative ways in which climate change can be not just adapted to, but also mitigated by, solar design thinking.

Notes

1 GERES began working on passive solar projects in Ladakh in the early 1980s, but no longer conducts fieldwork in India (Geres 2020).
2 Notably, the villages of both Alchi and Stok are north facing, and indicate a preference for the site beyond solar ideals.
3 Some villages lack access to roads altogether, making the transportation costs associated with imported fuel almost insurmountable.

References

Daultrey, Sally, and Reuben Gergan. 2011. "Living With Change: Adaptation and Innovation in Ladakh." *Climate Adaptation*. Available: www.yumpu.com/en/document/view/25089047/living-with-change-adaptation-and-innovation-in-our-planet.

Dutta, Sanjay. 2019. "Ladakh Will Soon Be Home to World's Largest Solar Plant." *Times of India*, January 13, 2019. https://timesofindia.indiatimes.com/india/ladakh-will-soon-be-home-to-worlds-largest-solar-plant/articleshow/67507422.cms.

Ferrari, Edoardo Paolo. 2018. *High Altitude Houses: Vernacular Architecture of Ladakh*. Florence, Italy: Didapress.

Geres. 2020. "Where We Work – Geres." January 1, 2020. www.geres.eu/en/our-actions/countries-of-intervention/.

Hales, Carolyn. 1986. "The Ladakh Project." *Cultural Survival Quarterly Magazine*, September 1986. www.culturalsurvival.org/publications/cultural-survival-quarterly/ladakh-project.

Heath, Cai. 2015. "Climate-Friendly Development: Analysing Relationships between Community, Society and Government on Sustainable Technology Projects." *Ladakh Studies* 32 (January): 18–35.

India Today. 1994. "With Trombe Walls and Direct Gain Technology, Ladakh Experiences First Solar Revolution." India Today. November 30, 1994. www.indiatoday.in/magazine/special-report/story/19941130-with-trombe-walls-direct-gain-technology-ladakh-experiences-first-solar-revolution-810298-1994-11-30.

Jacobsen, Arne. 2000. "Rural Electrification Using Photovoltaics in Ladakh, India." In *Methodological and Technological Issues in Technology Transfer*, edited by B Metz, OR Davidson, J Martens, SNM Rooijen, and LVW McGroy, 409. Cambridge: Cambridge University Press. http://schatzcenter.org/docs/Jacobson-1999-CaseStudy14RuralElectrificationLadakh.pdf.

Joshi, Neelakshi. 2013. "Efforts to Resurrect and Adapt Earth Building and Passive Solar Techniques in Ladakh, India." In *Vernacular Heritage and Earthen*

Architecture, edited by Mariana Correia, Gilberto Carlos, and Sandra Rocha, 611–16. New York: CRC Press.

Mingle, Jonathan. 2015. *Fire and Ice: Soot, Solidarity, and Survival on the Roof of the World*. 1st edition. New York: St. Martin's Press.

Norberg-Hodge, Helena. 1991. *Ancient Futures : Learning from Ladakh*. San Francisco: Sierra Club Books.

Prakash, Sanjay, ed. 1991. *Solar Architecture and Earth Construction in the Northwest Himalaya*. Sustainable Development Series 5. New Delhi: Har-Anand Publications in association with Vikas Pub. House.

Stauffer, Vincent. 2004. *Solar Greenhouses for the Trans-Himalayas: A Construction Manual*. Kathmandu : Aubagne, France: International Centre for Integrated Mountain Development ; Renewable Energy and Environment Group.

Tonge, Peter. 1982. "THE TROMBE WALL; A Good Idea Wherever It Travels." *Christian Science Monitor*, July 21, 1982. www.csmonitor.com/1982/0721/072140.html.

Wangchuk, Rinchen Norbu. October 30, 2019. "23 GW of Potential and Growing: How Ladakh Plans to Lead India's Solar Charge!" *The Better India*. Accessed at: https://www.thebetterindia.com/201532/how-much-solar-power-india-rates-cost-ladakh-industry-information/.

10 Greenhouses

While the trans-Himalayan region has long sustained village agricultural communities, this work has always required assiduous resource allocation and farming practices specifically tailored to the region. These two traditional farming approaches continue to be called for under the increasing pressures of climate change, and any new agricultural directions will also need to find an appropriate fit within the physical, social, political, and economic frameworks of the region. As climate change creates new environmental conditions in Ladakh, both challenges and opportunities will emerge to shape farming practices. The region's agricultural makeup will necessarily transition to support shifting needs and opportunities, because the anticipatory agricultural landscapes of the future cannot be modeled on the conditions of the past.

In addition to the many environmental changes taking place in Ladakh, livelihoods are also becoming more diverse. Many Ladakhis have followed new opportunities for education and employment that lead away from full-time farming, and those paths can dislocate individuals from family farmlands, or leave households with fewer available laborers. Demographic shifts have changed household makeup, and it is not uncommon for one or more individuals to find employment outside of their village, often for months at a time. At the same time, people who remain in villages to tend crops and livestock may seek out alternative farming practices that support dietary changes, emerging trade opportunities, or reduced household workloads. Social and cultural conceptions of village farming, along with perceived socio-economic status, are also adjusting to reflect interests outside of agriculture.

Meanwhile, political forces also appear to be steering Ladakhi agriculture in new directions. During the course of the past several decades the government has provided residents with a state ration that enables households to supplement local provisioning with basic staples brought in from the subcontinent. Increasingly, many households can afford to scale back their subsistence agricultural practices, using money earned elsewhere, as well as the ration, to supplement basic caloric needs. This new dependence upon external sources, in conjunction with changes to the economic, social and climactic makeup of Ladakh, ushers in new avenues for food insecurity.

During the late fall, winter and early spring, on-site growing in Ladakhi villages is severely limited by the weather. At this time food cannot be delivered to communities by road, as high mountain passes are closed by snow. In major towns, such as Leh, fresh and imported food may be brought in by airplane, although these products do not reach rural villages. The expense and high environmental costs of importing food, by either truck or plane, makes a case for locally available food.

A passive solar greenhouse enables farmers and growers to take advantage of the sun that heats this region during all of the months of the year, by enclosing an interior space where ambient temperatures will also allow for year-round agricultural processes. Greenhouses provide sheltered interior spaces for food, biofuel, compost and flower production. These spaces can be built to harness passive solar warmth, functioning without additional fuel sources. In this way, the indoor climate of the greenhouse creates warmer growing conditions, effectively extending the growing season and allowing for food production even in winter months.

Greenhouses facilitate the potential for off-season farming, low-water use, pest and predator protection, and a better variety of crops. Such environments could offer new space for winter vegetables, as well as places to rear livestock like poultry. This design solution amounts to a low-risk, high reward project that can scale up across the region in the face of climate change pressures.

The design solution provided by greenhouses falls somewhere between economic opportunism and necessary food security in contemporary Ladakh. Greenhouses provide additional earning opportunities for growers by facilitating more diverse crops, with relative protection from freezes, snowfall, and pests. In Ladakh, summer is the primary season for agricultural production, because this is the only time when ambient temperatures will allow for most outdoor plant growth. Finally, the benefits and opportunities provided by greenhouses suggest that this work could be done in conjunction with existing field farming, or in addition to another enterprise, at minimal risk, investment, and cost.

Background

The Ladakhi diet was once extremely limited, due to the region's challenging climate for food production and the difficulty of transporting food products along high-altitude trade routes. Until recently, imported tea and spices supplemented staples of locally sourced peas, barley, and wheat (Crook and Osmaston 1994). Minimal meat and milk consumption, sourced from yaks, and later, cattle, goats, and sheep, rounded out this diet. As roads and vehicles gradually wound through the region, access to imported foods has grown. Today, a diverse array of staples are readily available in many parts of Ladakh, along with international brands of chips, soda, and other processed foods. The flavors and food preferences in many villages reflect local agricultural offerings combined with influences from the rest of India and beyond. Bridging the traditional diet with external influences, Ladakhis have absorbed new social practices around eating and shared meals, adopted modern types of cooking devices (such as microwaves) and expanded daily meals to reflect the wide array of food available.

The government ration system, called the Public Distribution System (PDS), may be the single greatest contributing factor in the shift away from a traditional Ladakhi diet. It was created by the British colonial government in the 1940s in an effort to reduce poverty, as well as to curb price gouging and provide Indian citizens with a safety net in the face of food shortages (Dame and Nüsser 2011). While this service effectively reduced food insecurity across India, particularly among the poor, it also created new dependencies for Ladakhi village households. Dame and Nüsser note that this reliance has cost Ladakhis some of their hard-won self-sufficiency, where "Dependency on subsidies renders the population vulnerable should the national government decide on adjustment of the programme or major reforms" (Dame and Nüsser 2011, 191).

In aggregation, all of these factors paint a picture of transition in Ladakh, and one in which robust food security becomes gradually diminished. Malnutrition is now a real concern for residents in Ladakh, although the region does not exhibit undernutrition or starvation (Dame and Nüsser 2011). In their 2011 study on agricultural transitioning in Leh, scholars Juliane Dame and Marcus Nüsser recommend that "Enhanced preservation and storage capacities, a growth in off-seasonal vegetable production, income opportunities and the advancement of education and nutritional knowledge should be priorities for supporting household strategies" (Dame and Nüsser 2011, 191). Climate-adaptive design responses for agricultural supports have an ability to dovetail with this call, strengthening Ladakhi food sovereignty in the process. As a single strategy, greenhouse farming offers village households an opportunity to extend the growing season, with increases to overall agricultural production that could ultimately bolster the diet under the ration while producing new economic opportunities.

In a climate that effectively limits farmers to a single growing season each year, and a context where few storage facilities exist to keep fresh vegetables, micronutrient deficiencies are prevalent (Dame and Nüsser 2011). If Ladakhis had access to a more varied diet throughout the year, these nutritional problems could be reduced or even eliminated. Dame and Nüsser cite a "pronounced seasonality of dietary patterns" (Dame and Nüsser 2011, 184), wherein vegetables and imported foods are available to villagers in the summer months, but for a long multi-month winter when both roads into Leh District close, dietary consumption reverts to the local crops that store well in this cold context. Greenhouses help to create a local supply of fresh winter and shoulder season produce, and therefore support more diversified diets during these months.

In addition to better health outcomes, greenhouses also facilitate income generation: they create new markets, and may provide new opportunities to subsistence farmers who also want to participate in the cash economy. The warm interior climates of greenhouse can support introduced summer plants and much more diverse selection of fruits and vegetables across the arc of the year. If new sources of fresh food can be grown in Ladakh, the burgeoning tourism industry and the troops who live on the strategic military bases throughout the region represent potential markets. But Dame and Nüsser suggest that this opportunity is currently underused, as only a "few households specialised in vegetable production" (Dame and Nüsser 2011, 188). Fresh vegetables produced in greenhouses can be sold at the bazaar in Leh, where individuals and guesthouse staff are willing to pay premiums in late spring and early summer (before the roads open to allow produce to come in from other parts of India) and directly to restaurant and army kitchens.

Greenhouses have been incorporated into Ladakhi households over the past few decades, with a range of outcomes. Both governmental horticultural programs and NGOs played a part in the widespread adoption of various greenhouse models, and the bulk of these interventions were introduced into Ladakhi communities rather than built up from within. Groupe Energies Renouvelables, Environnement et Solidarités (GERES) started working in the region in 1982, and at this time Ladakh Ecological Development Group (LEDeG) had piloted several greenhouse designs as well (Stauffer 2004). Ladakh Health and Environment Organization (LEHO) is credited as the first NGO to build a widely adopted greenhouse model, by training builders and developing the design (Stauffer 2004).

The government initiated an enormous greenhouse building effort in the 1990s, with a model that used polythene sheets in contrast to other earlier versions using glass. Although more than 15,000 greenhouses were produced, this early model did not create space that was warm enough to support winter production in very cold climates (Triquet et al. 2009). While the project helped to seed the idea of greenhouse use in Ladakh, and popularize the practice of greenhouse growing, only 10 percent of these structures were in use a decade later (Triquet et al. 2009).

Today a variety of greenhouse designs exist, but the LEHO passive solar greenhouse is perhaps the most popular, and it remains visibly predominant across Ladakh. This success is due in part because it is specifically tailored to the needs of families in the region and in part because it has, over the years, been heavily promoted by a number of NGOs.

How it works

The LEHO passive solar greenhouse incorporates four different design requirements into the functioning of a single space over the course of a day. Greenhouses must orient to the sun, store the radiation during the day, radiate heat out into the space during the night, and minimize heat loss throughout this 24-hour period (Stauffer 2004, 9). To do this, greenhouses are almost always oriented along an east–west axis, with the high rear wall on the north side. Transparent materials face south, which facilitates the collection of solar radiation.

LEHO has specific recommendations for the placement of new greenhouse buildings. Their guides suggest that natural shading needs to not interfere with the solar access of the structure, so that sunrise occurs before 9:30 am in the winter and sunset after 3:00 pm (Stauffer 2004, 14). Water needs to be readily accessible in winter (when many streams freeze) as well as in the spring, summer and fall. They have found that flowing water should be no less than 600 feet in the winter, and 300 feet in the other seasons (Stauffer 2004, 14).[1] Overall, the site should be flat, dry, earthen, and south facing, with an interior size of roughly 28 feet by 15 feet (Stauffer 2004).

Heavy masonry materials must be used to build the greenhouse frame, typically on east, west and north sides of the building, in order to store and radiate heat throughout the course of the day. These walls can be buried into existing hillsides, connected to another building, or built up out of the ground. An outer layer is load bearing, while an inner layer is designed to store heat. Insulation made up of straw, grasses, or other lofty natural materials should be placed inside the cavities within these walls. Dark masonry walls absorb heat better than earth tone, so if possible, interior greenhouse walls should be painted black.

The LEHO model recommends digging down 6 inches into the ground so that the floor of the greenhouse lies below the surrounding terrain. This helps to insulate the space, which can be further improved by adding a layer of dung. In a report for GERES on the LEHO greenhouse model, author Vincent Stauffer notes that this shallow trench captures carbon dioxide, which can improve plant growth (Stauffer 2004, 13).

A door needs to be placed on one side of the greenhouse (usually a leeward east or west side), and this can be sized just large enough to enable growers to move themselves and their plants in and out. The interior space can become overheated or humid during the day, and so hand-operated vents in the roof are also important additions. These vents enable users to manually adjust the interior temperature, allowing for their thermal comfort while working in the greenhouse, and reducing problems associated with mold, rot and disease.

Finally, if needed, a roof on the north side of the building should be made of material with insulation to reduce heat loss. In Ladakh this roof should be angled at 35 degrees, to reduce the overall interior space of the greenhouse, and to allow for optimal solar radiation on the interior surface during winter months. Roof angles should either curve towards the south, or be set at two different angles. The front section needs to be 50 degrees or more, to capture solar radiation in early morning and late evening. The upper section on the south side is set at 25 degrees, to enable mid-day radiation and to allow snow to fall off using gravity (Stauffer 2004, 12). Taken together, these prescriptive recommendations for greenhouse construction constitute the typical LEHO greenhouse model, although the organization and its affiliates have adjusted various factors within this prototype to accommodate different sites, clients and growing purposes.

Materials

Many of the NGO-sponsored solar greenhouses provide cheap plastic sheeting to cover a lightweight roof structure, and these buildings can be seen in various states of disrepair throughout Ladakh (Figure 10.1). While sheeting is affordable and easy to transport to a remote site, it lacks the durability that would enable the greenhouse to become a permanent household fixture. Plastic sheeting has a tendency to fall apart during the course of a single season, and at minimum it needs to be regularly replaced every few years.

While this bare-bones greenhouse model remains the norm in Ladakh, occasionally materials are upgraded to reflect a desire for permanence or performance. LEHO, for instance, widely circulated their most affordable greenhouse model using polythene sheeting. But throughout Ladakh other LEHO-branded greenhouses are also visible. In the past, they have imported polygal sheets from the Indian subcontinent, offered at heavily subsidized rates to participating households. While these materials have relatively high upfront costs, they have a much higher likelihood of performing over time, helping to ensure years of access to a greenhouse and conserve human energy.

In addition to these subsidized models for individual households, new government programs and other major organizations have also invested in high-end, permanent greenhouse buildings (Figures 10.2, 10.3, and 10.4). As in the fortified LEHO model, these structures are made up of rigid polygal or glass panels that are both stronger and more permanent. They can have fans, electric lighting, and other technologies. These greenhouse projects tend to be built on larger plots of land with a mission to provide food for an organization, or for research, rather than for individual household consumption.

LEHO's greenhouse construction guidebook specifies the incorporation of locally sourced materials wherever possible. With the exceptions of fasteners, equipment and polygal or polythene sheeting, the majority of greenhouses in Ladakh reflect adherence to this mandate. Poplar wood poles and beams, talu (willow branch) roofing, and mud bricks can all be hand harvested and formed within villages. In this way, the upfront cost for the structure is minimized, and local growers can independently harvest, build, and maintain the product over time.

Figure 10.1 Plastic sheeting stretched over a greenhouse lacks durability in Ladakh.

Figure 10.2 Signage on the road outside of Leh advertises greenhouse research facilities.

Figure 10.3 Research greenhouse space, behind a fence on land not otherwise used for agriculture in Ladakh.

Figure 10.4 Inside of the greenhouse in Padum, a variety of crops are grown.

Financing

The LEHO passive solar greenhouse model strives to enable affordability by specifying the use of predominantly local materials. One metric for affordability, with this system, is that the construction of a single greenhouse should be paid for by selling well-managed produce from the greenhouse, within a period of three years (Stauffer 2004, 5). In addition to this self-financed model, a variety of arrangements over the years have provided loans, gifts, and subsidies to households, organizations, and individuals to promote greenhouse development.

In their study of program participants, Dame and Nüsser note that the application of greenhouses in Ladakh requires an initial outlay of money, which necessarily excludes the most economically vulnerable households. In this way, even subsidized self-help programs may serve to create wealth in families that can afford to participate, while limiting opportunities in the households who need this support the most. Dame and Nüsser found that "Comparatively well-off households, that can afford the cash investment, especially benefitted from the (greenhouse) scheme. As financial assets are indispensable for programme participation, a certain number of already more vulnerable households were excluded" (Dame and Nüsser 2011, 187).

Benefits

Greenhouses improve food security and may promote additional healthy choices in the Ladakhi diet. These buildings provide a space to grow produce in cold winter months, which can allow for the production of winter vegetables and thus increase access to food sources even while other avenues disappear. These new sources of nutrients can also be amendments to the limited winter diet: according to Vincent Stauffer, spinach, carrots, and onions can all be grown in winter greenhouses in Ladakh. Year-round growing, in the extreme climate of Ladakh, provides a real benefit for village households, particularly in those areas that become cut off to external trade sources in the winter months.

Greenhouses also support efficient and affordable growing conditions. The covered structures reduce evaporation, and so crops require less water than they would in conventional outdoor growing situations. Free year-round heat from the sun creates a pleasant space to work, especially in the cold winter months, when ambient temperatures often dip below 0 degrees Celsius. In most climates, greenhouses enable seedling production in the spring to jumpstart early planting, the production of introduced (exotic) vegetables in the summer to augment the local diet and to support tourism and army troops, and additional crop production in the fall.[2] In very cold climates, greenhouses' fresh vegetable production is typically around $0.8\,kg/m^2$, and as much as $1.4\,kg/m^2$ in cold climates (Stauffer 2004). According to Vincent Stauffer, in extremely cold climates greenhouses can provide, at minimum, leafy vegetables in the winter months, and seedlings and root vegetables in the spring, and root vegetables in the fall. From tomatoes to root vegetables, the passive solar greenhouse design allows growers to increase the quantity, diversity, and availability of fresh produce.

This additional growing capacity also enables Ladakhis to be opportunistic about agricultural profitability. Beyond serving the dietary needs of individual households, passive solar spaces can create a real source of income for locals. Taking advantage of their remote location, greenhouse growers can cater to the local tourism industry, as well as the presence of army bases and government outposts. By producing greenhouse products in Ladakh, particularly during the seasons in which road serving the area are closed, farmers can be paid well for agricultural efforts. The development of greenhouse enterprises may even provide targeted opportunities for women, who typically handle much of the agriculture and selling in Ladakh.

Limitations

Greenhouses may provide numerous benefits to growers, but their overall impact, at a household level, likely also has an upper limit. Site conditions, labor availability, and material costs constrain the size of a greenhouse enclosure. Most greenhouse crops require daily management for watering and to prevent overheating. Perhaps the greatest challenge to the solar greenhouse is that it requires specific conditions in order to work at all, and so if a year-round water source is not available nearby, or if a mountain casts a shadow on a potential site, the project will not be viable.

Materials also become a limiting factor, as glazing cannot be locally sourced and costs a great deal in the remote villages of Ladakh. Added to this constraint, the lifespan and cost of outsourced materials may deter implementation. Material choices also impact performance. For instance, plastic sheeting is far less conducive than glass in producing the greenhouse effect, whereby heat radiates into the greenhouse and is stuck inside, leaving instead through conduction. Compared to glass, polythene is 50 percent less effective as a material choice, although it is considerably more affordable and much easier to transport in Ladakh (Stauffer 2004, 10).

Moving forward

In addition to the standard freestanding greenhouse promoted by LEHO and other NGO partners, unique versions of the greenhouse design idea can be found across Ladakh. Last year, a social enterprise called Agrow created the first greenhouse made up entirely of waste plastic bottles, plastic trash, and sand, in the village of Chemrey (Karelia 2019). This approach provides multiple benefits: removing plastic waste from landfills, improving the strength and effectiveness of the standard greenhouse envelope, and promoting vegetable cultivation to reduce food miles.

Another promising climate-adaptive design idea is the attached greenhouse, which acts not only as a space that can be used for food production proximate to the house, but also becomes an additional building armature to improve the functioning of the interior home environment. These spaces are useful in the winter and shoulder seasons, but like the *shelkang* (solarium) in passive solar architecture, can generate excessive heat in the summer months.

When built as a solar veranda, greenhouse solariums can add livable space to the house and also increase the perceived value of that space. These types of spaces work very well when built upon a more permanent wall and foundation, with structural wood framing and operable or removable glass panes. In this configuration, indoor gardening space also produces a thick thermal barrier for the house itself, acting as a sunroom that enables both growing and thermal comfort. NGOs have helped to disseminate greenhouse designs and promote the idea, and examples of attached greenhouse/sunrooms can be found across Ladakh today.

Conclusion

As a climate-adaptive design solution, greenhouses tend to use appropriate technologies and approaches for the region of Ladakh. The structures incorporate vernacular building methods, but layer over technological recommendations from outside experts to maximize the overall efficacy of the building. These spaces vary greatly in terms of specific shape and function, but as a strategy for bolstering food production in Ladakh, they have, on the whole, a proven record of performance.

Despite this long record of experience and relative ubiquity, the solar greenhouse concept has room for improvement. Typical greenhouse systems are restrained to say the least, in terms of goals and exposure to risk. Perhaps because of this it is also one of the most used, appreciated, and adopted strategies in Ladakh. Against this backdrop, there is an opportunity to test other types of wild ideas, including more technologically advanced solutions and design ideas that incorporate specific alterations for individual stakeholders.

One logical area for growth can be found in crop production. Are there high-value crops currently missing in this context, such as tea varieties, spices, or medicinal plants? Might animals claim habitation space in a greenhouse? What about fruit and nut trees? Current greenhouses tend to produce limited varieties of vegetables and fruits, and, in the coming years, more diverse crops could help to leverage the value of these spaces.

In addition to reconsidering production, Ladakhi greenhouses could benefit from reimagining the relationship between buildings. Could there be additional, opportunistic spatial adjacencies among buildings on a single lot? There may be an opportunity to reinforce connections with architecture, whether that involves co-locating greenhouses with residences, barns, or composting toilets. Currently the massing provided by the greenhouse shell is used primarily for the thermal needs of that space, rather than for the containment of animals, or indoor living space, or other types of protection from pests and predators. By linking greenhouse design to architectural form, growers may be able take advantage of additional synergies produced by co-location.

Against the backdrop of a shifting Ladakhi workforce, one important question is to ask is: Do households actually need more agricultural space? Are there better opportunities for people through education, technology, and tourism? Perhaps there is good reason to move away from subsistence agricultural livelihoods, especially if the climate itself becomes too unreliable to depend upon for growing. On the other hand, new types of agricultural space may improve opportunities for individual households in the context of climate change and the shifting needs under different demographic trends.

Even if fewer people exclusively farm in the future, there is merit to providing some agriculture products on site in Ladakh. Indeed, the climate is challenging, but so too are the roads, and regular access to the kitchens at the ends of those roads. People in Ladakh need to have reserves in place, for times of crisis, as well as for the daily desire for fresh, local products.

After all, the large-scale farming that supports India's ration system also has the potential to threaten Ladakhi self-sufficiency. Political scientist and anthropologist James C. Scott has written extensively about the dangers of exchanging small-scale local farming production for the large and remote enterprises created under Green Revolution principles. He notes that "Modern, industrial, scientific farming, which is characterized by mono-cropping, mechanization, hybrids, the use of fertilizers and pesticides, and capital intensiveness, has brought about a level of standardization into agriculture that is without historical precedent" (Scott 1998, 266), and while this type of farming is not currently practiced in Ladakh, it has the potential to displace Ladakhi food production through the ration and trade. Moreover, Scott asserts that:

> The simple, "production and profit" model of agricultural extension and agricultural research has failed in important ways to represent the complex, supple, negotiated objectives of real farmers and their communities. That model has also failed to represent the space in which farmers plant crops – its microclimates, its moisture and water improvement, its microrelief, and its local biotic history. Unable to effectively represent the profusion and complexity of real farms in real fields, high-modernist agriculture has often succeeded in radically simplifying those farms and fields so they can be more directly apprehended, controlled, and managed.
>
> (Scott 1998, 262)

Scott asserts that traditional agricultural practices can be an effective way to hold power, and to resist state control.[3] After all, reliance on the Indian ration (PDS) creates a tether to the state that can be viewed as a double-edged sword. Ladakhis who become dependent upon the government for staples may inadvertently also cede their ability to make decisions about what, when, and how they eat. In contrast, farming communities who continue to bolster their own self-sufficiency through climate-adaptive agricultural methods retain relative autonomy from the state. In this way, growing is inextricably linked to self-governance, and the act of self-provisioning becomes political.

Landscape architect Kenneth Helphand also makes a connection between the acts of agricultural production and political defiance. In political, economic, social, and cultural terms, gardening can be a form of resistance. But Helphand also notes that defiant gardening may ultimately also be contextual.

> Defiance may also be directed against environmental conditions, the extremes of climate, difficult topography, lack of soil, water or even plants. Gardens in extreme environments of heat, cold, elevation or drought accentuate the material origins of the garden and its forms.
>
> (Helphand 1997, 106)

The greenhouses born out of Ladakh's difficult climate showcase a version of defiant gardening that helps to shore up food security under environmental stress but in the process, also bolsters social, cultural, political and economic autonomy. The result of these efforts is a form of agriculture that directly helps Ladakhi communities to transition to new modes of self-provisioning and economic participation under climate change.

Notes

1 Longer distances create disincentives for use because of the difficulty of carrying water to crops. Winter months require less watering, and are more likely to diminish stream availability, so the distance can be a bit greater (Stauffer 2004, 14).
2 For the purposes of greenhouse growing conditions in Ladakh, climates can be divided into the categories of cold (less than −10 °C), very cold (between −10 and −15 °C) and extremely cold (greater than −15 °C) (Stauffer 2004, 4).
3 "For the architects of state space, any substantial move from wet rice at the core toward foraging at the remote periphery is a threat to the manpower and foodstuffs underwriting state power" (Scott 2009, 185).

References

Crook, John, and Henry Osmaston. 1994. *Himalayan Buddhist Villages: Environment, Resources, Society and Religious Life in Zangskar, Ladakh*. Bristol: University of Bristol.
Dame, Juliane, and Marcus Nüsser. 2011. "Food Security in High Mountain Regions: Agricultural Production and the Impact of Food Subsidies in Ladakh, Northern India." *Food Security* 3 (2): 179–94. https://doi.org/10.1007/s12571-011-0127-2.
Helphand, Kenneth. 1997. *Defiant Gardens*. San Antonio: Trinity University Press. http://tupress.org/books/defiant-gardens.
Karelia, Gopi. 2019. "Made of Plastic Waste, This Unique Greenhouse Will Help Ladakh Farmers Grow Veggies All Year!" *The Better India*, September 25, 2019.
Scott, James C. 1998. *Seeing Like a State: How Certain Schemes to Improve the Human Condition Have Failed*. New Haven: Yale University Press.
Scott, James C. 2009. *The Art of Not Being Governed: An Anarchist History of Upland Southeast Asia*. New Haven: Yale University Press.
Stauffer, Vincent. 2004. *Solar Greenhouses for the Trans-Himalayas: A Construction Manual*. Kathmandu: Aubagne, France: International Centre for Integrated Mountain Development; Renewable Energy and Environment Group.
Triquet, Marion, Tashi Tokmat, Konchok Dorjai, and Vincent Stauffer. 2009. "Passive Solar Greenhouse in Ladakh: A Path to Income Generation and Livelihood Improvement." Geres and LEHO. www.solaripedia.com/files/1087.pdf.

11 Reservoirs and canals

Zings, *yura*, and *kuhls*

Historically, Ladakhi villages have relied on low-tech, time-tested water management systems to funnel high glacial and snowfield meltwater down to low-lying croplands. The traditional canals, water diversion mechanisms, and organizational systems that enable equitable sharing, have, for centuries, formed the backbone of subsistence agricultural practices. As scholar Janet Rizvi notes "The scantiness of rainfall makes Ladakhi agriculture totally dependent on irrigation; but all those sources that lie within the scope of traditional technology have been tapped" (Rizvi 1998, 188). Indeed, other scholars also suggest that traditional Ladakhi irrigation practices have already been well optimized as a response to the cold desert environment (Crook and Osmaston 1994; Norberg-Hodge 1991; Rizvi 1998).

In recent years, however, the changing weather patterns and diminishing supply of meltwater brought about by climate change has inspired improvisation within these traditional systems. Even within a system of judicious water management, new efficiencies may be found to better manage the flow of surface meltwater, to redirect unused water, and to stockpile irrigation reserves for deferred agricultural use. By building upon the foundation of traditional water management systems, and overlaying assorted adaptive design interventions, the overall system efficiency may be further improved.

A number of villages have sought to improve their water management infrastructure by requesting technical assistance from local NGOs and state government agencies. In other cases, individual households simply try new ideas or adjust a popular design to fit their own needs. Several relatively innovative new practices have been adopted by Ladakhi households and even entire villages, typically in the form of a design intervention that tweaks an existing infrastructural system to bolster water functioning.

These climate-adaptive solutions have been developed and deployed on a case-by-case basis, often as a result of stakeholder involvement, such as consensus decision-making techniques or communitarian builds involving sweat equity. As such, the disparate responses represent a portfolio of related design interventions that might be deployed as needed in a wide variety of water scarce environments. This body of adaptive management practices reinforce the systems already in place in Ladakh, improve the functioning of those systems, and in so doing represent an effective small-scale planning solution for the social, cultural, environmental, economic, and political implications of this unique place.

Background

Skillful irrigation was said to have originated in the area in the tenth century (Bell 1928); and scholars have noted that "Without irrigation vegetable farming is out of the question in Ladakh" (Acharya et al. 2012, 312). In the cold arid landscapes of Ladakh, agriculture is both intensive and relatively productive, made possible only through judicious water stewardship. Most of the water used for irrigation comes from above: in addition to glacial meltwater, winter precipitation produces a supply of snow that will melt during the course of the following growing season.

Water is primarily captured by villages nestled in the creases of mountains, as it melts from glaciers and snowfields above 17,000 feet and descends, through tributaries, past farms and fields. In these communities water is shared as a part of the commons, equitably distributed through a variety of "rules in use" (Ostrom 1990). According to scholars Sally Daultrey and Reuben Gergan, "The main source of irrigation in Ladakh is surface water, with approximately 10,901 hectares of land around the tributaries to the Indus irrigated by a sophisticated and carefully managed series of small, hand-built mud canals, which make effective use of seasonal run-off from melting snow and ice at high altitudes" (Daultrey and Gergan 2011, 5).

In addition to the ancient practice of harnessing meltwater using gravity-fed irrigation systems, a number of farming households benefit from their proximity adjacent to the Indus or Zanskar Rivers and are able to draw from these sources for their agricultural use. Because low-lying rivers have considerable capacity, the rotational systems used to equitably manage water among households are not necessary. Instead, river water is simply directed through embankments and inundation channels as needed.

Springs located at various public taps within villages provide a source of water for households; typically small-scale uses such as drinking, cooking and cleaning (Figure 11.1). A final form of water access is found in wells; like springs these wells can be used to provide drinking and washing water for households but typically do not provide a source of irrigation for agriculture. Wells have been heavily tapped in communities supported by the tourism industry, where on-demand water feeds new fixtures such as flush toilets and showers. This additional use impacts the aquifer, and represents a fundamental shift away from traditional Ladakhi water management practices. While all households once depended upon meltwater for farming, increasingly enterprising families now source water from underground wells to support tourism.

Nevertheless, irrigation water remains an important part of the equation for farming households. In every village families have a specific share

Figure 11.1 A spring line carries water down in elevation to a village below.

of water, or water rights, which have been set by their forefathers. Ancient maps and papers, often held by monasteries or in district government offices predating British rule, outline these age-old agreements and under-standings (Sharma and Bharat 2017). If a household doesn't have the right to access a village *tokpo* (gravity-fed stream) then they may be entirely dependent upon springs or tertiary runnels for their water use.

Inter-village agreements also establish water rights, when a *tokpo* is shared among more than one village, the upstream village may enjoy a natural advantage, and so agreements on water allocation ensure equitable distribution. When one upper village closes a *yura* (dispersion canal) to let the water pass by fields to feed a community sited lower in the drainage, it is called *pabcha*. *Pabcha* not only is the agreement for the allocation of water, but also any stipulation for an exchange (such as the delivery of animals or alcohol). The *pabcha* system has been traced back to written references in 1571, with records identifying water rights between villages, and royal orders dictating a certain number of days and nights of watering (Dawa, Dana, and Namgyal 2000, 245). During the course of intervening years, various Ladakhi kings and subsequent leaders have weighed in on these agreements, making micro-adjustments when necessary, or helping to resolve disputes between villagers (Sharma and Bharat 2017). Despite the many transitions in rule and leadership over the years, these agreements have remained largely intact. The Jammu kingdom, British officials, and even the current Indian government have acknowledged and then largely defended the extant system for water access over the years.

The complex arrangements for water rights has yielded a similarly complicated and structured system for water management. Specific agreements dictate the exact timing of the water stoppage, and redirecting, as well as the people who are invited to participate (from a monastery or gompa), affected families, elected leaders and shifting guards. *Churpons*, village water overseers are also identified at various intervals to manage this shared system of water access. Dawa et al. (2000) and Kim Gutschow (1993) have conducted thorough studies of water rights in various regions of Ladakh and document the complexity of water infrastructure in practice (Dawa, Dana, and Namgyal 2000; Gutschow 1993).

Traditional water systems

In Ladakhi, ice water, or *kangs-chhu*, is the meltwater captured primarily for agricultural use. *Tokpo* (or *togpo*) are the streams that carry the meltwater down to villages, and are typically fed by glaciers and snowfields. The *tokpo* shifts in volume over the course of the year, depending on melt from solar insolation and other environmental factors.

Yura are the channels that drain water away from the *tokpo*. These are hand dug, and gravity fed; they serve to transport water into specific fields or reservoirs. They can be made out of earth, and indeed most are simply a carved runnel in the land. But *yuras* can also be improved to reduce the pervious qualities of the channel and to stop it up when needed. To assist with management and communication, there is a fairly complex system of organizing and naming various channels. For instance, the *ma-yur* is mother channel, and will often be lined with clay, like a dyke (Angchok and Singh 2006, 397). Within *yura* there are *raks*, which are dry stone bunds, dykes or weirs, rka, gaps or sluices, and *rka-do*, rocks that can be used as stoppers.

It is not uncommon to see individuals using modern materials within these age-old systems. For instance, a channel might be stopped with some castoff socks left by trekkers, or a historic channel could be lined with plastics, metals or concrete. These micro-design interventions help the traditional water system to perform better, without causing a shift away from the original use.

How it works: the *zing*

Zings (or *dzings*) are large water reservoirs, usually located just above croplands in order to store meltwater for agricultural use. *Zings* are the ponds or reservoirs that hold water during spring, summer and fall, storing water for later use. These depressions can be carved out of the earth, lined with surrounding soils, or bolstered with dry stacked rock and even, in special cases, concrete. The reservoirs are typically constructed from masonry or bermed earth, and are then lined with concrete or mud. These tap into the gravity fed system of *yura*: At the upper end they will feature a gravity-fed inlet, while the lower side will have a drain and overflow gate. Water diverted into the *zings* can be controlled, both to direct it in and to release it into the fields. *Zings* are connected to the source of the meltwater and to the fields below by the same canals that have historically supported irrigation practices in Ladakhi villages for centuries (Figure 11.2).

These small ponds have a limited capacity, so they cannot provide extensive stores of water. However, they do enable irrigation flexibility, allowing downstream farmers to irrigate fields according to their own schedule. For instance, often fields are on a rotation of night or daytime watering, and instead of actively watering crops during the night, the field owners can divert the water to a *zing* to hold it for use the following day (Gutschow 1993). By storing water, even temporarily, above the sites where it will be used, *zings* ensure that all of the water available to farmers during a season will be used, day and night, thus reducing waste.

Zings also help farmers temper the diurnal swing of water availability. Because meltwater from high glaciers and snowfields is greatest in the afternoon, there can be a surplus of water available to farmers in the afternoons

Figure 11.2 A *zing* holds water for agricultural use, as a pond.

and evenings, with limited availability during the rest of the day. In order to ensure fairness and an equitable supply among households, *zings* act as storage space to enable individuals to divide and apportion this scarce resource more equitably.

While *zings* have been in use for centuries, they hold particular significance under shifting weather patterns, water scarcity, and unknowns of climate change. The use of a *zing* effectively creates a climate-adaptive buffer space – it is a small-scale example of a water storage tank that can be used. These systems have limitations as well. Due to their small holding capacity, the *zing* may only buy a farmer an additional day of irrigation, and they may not even accommodate the space needed to effectively contain water produced by major storm events.

Yura and *zings* are the most traditional, widespread and visible irrigation management ideas at play in the Ladakhi subsistence village landscape. Several other, smaller, and site-specific examples of water management design are also evident. In circumstances where *tokpo* disappear and go underground (likely filling the aquifer by infiltrating loose boulder or rocky soil, a workaround called a *yursar* or a *yursal* has been created. This captures the water in the *tokpo* where it is above ground, and rather than allowing this volume to run its course underground, diverts it into a special channel that directs water around the pervious landscape. In this way, the *yursal* or *yursar* hold the water above ground, and direct it using the same gravity fed *yura* lines, towards fields (Dawa, Dana, and Namgyal 2000, 243–4).

How it works: the *kuhl*

Kuhls, or canals, help to address one of the major challenges for today's farmers: the inefficiency of traditional irrigation infrastructure. Ancient channels linking upper glaciers to low-lying fields were carved out of the landscape over many hundreds of years; these are the mud, wooden, or stone lines carrying water along a path. These waterways lose a large amount of their volume to seepage; in one estimate 90 percent of the water routed above the village of Stongde, in the neighboring Zanskar valley, vanished before reaching fields (Crook and Osmaston 1994). Newer, high- and medium-tech *kuhls* reinforce these crumbling old irrigation lines, improving water access by reducing losses due to seepage and poor engineering.

Early *kuhls* were made of local materials; rocks create the mouth of the *kuhl* while the channel itself is a trapedozidal shape made of rocks and soil mortar, or earth. More recent attempts to improve these channels typically involved hardening this profile with a cement liner. According to Bhaskar Shrinivasulu Padigala, "A single *Kuhl* can irrigate an area of 80 to 400 ha through distributaries (40 cm × 50 cm width small channels for farms) or by flooding" (Padigala 2017, 642). *Kuhls* can run from 1 km to 15 or more km, diverting meltwater from high mountain streams to croplands via a channel. The *kuhl* is used to transport water across the land, sometimes

Figure 11.3 A reinforced *kuhl*, or channel.

diverting water from one uninhabited drainage over into another watershed by using gravity (Figure 11.3).

One of the largest irrigation interventions in Central Ladakh began in 1979, with the construction of the 43 km Igoo-Phey irrigation channel (Figure 11.4). This government-sponsored project has transformed the barren land along the southern side of the Upper Indus River into cultivatable croplands, which now host vegetable fields, animal pasturage, and trees for fruit and timber. According to Nüsser, Schmidt, and Dame, this single project increased cultivatable land in Leh District from 100 km² to 140 km² (Nüsser, Schmidt, and Dame 2012, 61). The project required considerable resources from the government, including the engineering and infrastructure that enables the canal to transition between various heights, cross waterways. The hardened banks of the canal today reflect the engineering prowess and toughness that characterizes many high modernist government schemes. While the project was started in 1979, the main channel did not reach substantial completion until 2000, and associated distributaries have also lagged behind schedule. Even today, much of the available land is not under cultivation due to an inability to allocate the land to individual farmers (Nüsser, Schmidt, and Dame 2012).

Compared to pre-existing systems, these new irrigation projects represent a more efficient and large-scale approach to water provisioning.

Figure 11.4 The Igoo-Phey *kuhl* pulls water from the river and transports it to agricultural holdings.

However, *kuhls* can still be considered supportive design interventions tailored to the landscapes of Ladakh. While they have been built with funding from government agencies (such as the Watershed Development Programme), they often draw construction labor from adjacent villages. Moreover, *kuhl* construction reinforces traditional Ladakhi cultural and social frameworks, by primarily hardening or improving existing water management systems that work within conventional agricultural concepts.

Financing

According to scholars Marcus Nüsser, Susanne Schmidt, and Juliane Dame,

One of the most prominent examples of intervention schemes is the National Watershed Development Program (NWDP 2009), which was initiated in Ladakh in 1995 for the purpose of integrated rural development, including expansion and improvement of irrigation infrastructure, agrarian innovations, and afforestation.

(Nüsser, Schmidt, and Dame 2012, 58)

Through this program NGOs can work to carry out water projects in Ladakh, and this has led to many hundreds of new projects sponsored by a variety of agencies. The small-scale village-based projects that have emerged under these programs have demonstrated a larger commitment to village participation, both in terms of the visioning of these projects and in terms of the building. Funding for this work comes from the *Watershed Development Project*, the Army's "Goodwill" Operation Sadbhavana program, and the Block Development Officer.

Benefits

Water infrastructure systems, such as the *kuhls* or the primary canals that direct meltwater to village croplands, have increased in size and scope over the past decade. *Kuhls* have been widened, improved with impervious membranes, regulator gates, and pipes, and stretched further than traditional materials would have otherwise allowed. They now have been engineered to leap over ridges, cross highways, and exploit topography to carry water across many miles to towns outside of a particular drainage.

Water management and dispersal has always been a critical component of the health and functioning of agricultural systems in Ladakh. Increasingly, design interventions have become part of broader attempts to conserve and reroute scarce water resources. Water-saving devices can be employed to help households manage drought. And traditional forms of water management are finding new utility when combined with new products, technologies, materials or design ideas. When paired with newer climate-adaptive design interventions, such as adding a *zing* to an artificial glacier system, both systems can be bolstered.

Limitations

Many of these water systems require management and oversight, to open and close valves in canals, to redirect water, to rebuild embankments that fail, or to dig out sediment that collects in *zings* or *yura* over time. There is a need to have individual oversight to ensure equitable allocation of water held as a form of the commons, with various elected positions for village representation and guarding.[1] Even within the context of this human-centric system, there may also be an opportunity to engage technology, perhaps to remotely open and close valves, or to provide surveillance. Even relatively simple technologies such as drip lines and sprinkler system are underused in current Ladakhi irrigation practice, and these devices could help to bolster water access or increase overall agricultural productivity.

The biggest challenge for the management of irrigation water comes from the materials used in various systems, rather than the design of the

Figure 11.5 Concrete used to reinforce a wall at Umlaat begins to fall apart.

systems themselves. Major seepage loss occurs in hand built earthen canals and retention ponds, which could simply be reinforced to boost the amount of water delivered to area farmers and households. However, plastic piping brought in in recent years degrades and breaks, and needs to be placed well below the frost line to prevent freezing. Other efforts to fortify retention walls and canal beds, with clay or concrete or other impervious surface treatment, are costly and require significant resources from external partners (Figures 11.5 and 11.6).

At one time the *yura*, *zing* and other irrigation interventions were built and managed by local village volunteers. This practice has since switched to some combination of government funding and NGO aid, both of which do not necessarily allow for regular, constant oversight. Villagers have become increasingly dependent on outside sources for assistance with major builds and with maintenance, which threatens their overall self-sufficiency. However, the heavy equipment needed to move large amounts of earth or to access imported materials such as plastic pipes or concrete, underscore the challenges of local adaptation through reinforced irrigation interventions.

Discussion

Unlike many other types of forward-thinking climate-adaptive design ideas, each of these water harvesting and management solutions supports

Figure 11.6 Plastic tarps have been applied to the *zing* above the village of Igoo, to manage bank erosion.

the existing agrarian lifestyles honed over centuries in Ladakhi villages. Substantial improvements to the system appear to primarily fortify traditional models. Rather than to reimagine the water access that will support village life in the years to come, these design interventions seek to improve the productivity of the existing system.

In assessing the economic implications of such interventions, one must first ask: Are they effective? Most hydrological models, especially those involving untested design interventions, are carried out over a series of years: indeed, one year wouldn't be adequately describe patterns or outlying factors. Current data on these water management strategies lacks this long-term view, and several of the interventions described in this chapter lack even the most rudimentary hydrological modeling. Thus, it is still too early to be able to tell whether the interventions work as anticipated, and to what extent they improve water access for cold desert farmers.

Anecdotally, however, in some studies these design interventions have been met with much interest and excitement from village farmers (Higgins 2012). The low cost of implementation, across the board, and the relative flexibility of each of these strategies suggests radically inclusive access for those who wish to employ these solutions. And in all cases, the water management strategies simply improve or shore up existing systems, so they represent a relatively low-risk response to climate-change adaptation.

While traditional surface water irrigation has supported life for many hundreds of years, structural flaws are evident in the broader planning system. One of the most difficult planning challenges for Ladakhi villages is that current irrigation methods don't allow for population growth; the system assumes a fixed number of individuals or farms. With better health and life expectancy through access to new medical practices, and with the dissolution of the ancient practices of primogeniture and polyandry, the number of households in the region will likely continue to grow. However, because the cultivable land in villages is fixed, individuals may need to seek employment elsewhere or import food for local consumption.

Likewise, the overall volume of water available in the watershed effectively places limits on the size and scale of an individual farmer's cropland; villagers must work together to develop an equitable division of irrigated fields. Ultimately, the volume of water coming from the high glaciers and seasonal snows may not be enough to support the crops grown below. After all, the intensive and highly productive agricultural practices of Ladakh rely heavily upon water resources, and are effectively limited by the region's supply of water.

Moving forward

The Ladakh Desert Development Programme (DDP) continues to maintain traditional canals and other irrigation projects in the region. While most of these infrastructural interventions already work, and satisfy many of the

Figure 11.7 Below the village of Nang, a *zing* is used for swimming, although this is an uncommon practice among locals.

needs of area farmers, they could be more efficient with small alterations. For instance, liners could be integrated in existing *zings*, *yura*, and *kuhls*.

Infrastructural landscapes could support multifunctional uses, and some of them already do. There is a very popular *zing* in the town of Leh that is used during the winter months for ice hockey and recreational ice skating. Other uses, such as recreational swimming, or fish production, could similarly be overlaid onto the reservoirs (Figure 11.7). Canals and various channels might be designed to produce a hiking trail, as the waterways are typically graded to move up to mountainous passes in a gradual manner.

Finally, there is an interesting new prototype wherein mustard or millet seeds are floated through the waterway of a *yura*, which leads through very loose soil, in order to start the planting process. The growth of the crops then serves to create a semi-pervious *yura* waterway, and allows the *yura* to become more effective in its primary role of transporting water. This idea is much talked about but relatively new to Ladakh, and according to Dawa et al., it has been piloted with some success by the Rural Engineering Wing of the Block Development Officer (Dawa, Dana, and Namgyal 2000, 251).

Conclusion

The irrigation resources in Ladakh have a long history of careful management, evidenced by ancient village water rights and rules in distribution and use. However, the structure of these management techniques ranges from local decision-making in one household or village, to much larger initiatives sponsored by the Hill Council, NGOs, and the government of India. This variety of actors produces a diversity in methods as well, with design interventions that range from gravity-fed water channels dug out of the earth to large regional projects such as canals. According to Nüsser et al., "A better understanding of how local irrigation and resource management systems are integrated in the broader mountain development process requires a consideration of political power and economic exchange relations between place-based dwellers and external actors" (Nüsser, Schmidt, and Dame 2012, 51).

The changing expectations about who and how water is managed reflects an agricultural society in transition. While it may highlight shifting environmental conditions, it is also connected to additional forms of provisioning, evolving economic opportunities, and new social or cultural ideals. Infrastructure that was once constructed and managed exclusively by village stakeholders has become increasingly the domain of government program officers, NGOs, or even hired non-local laborers. When villagers do participate in this work, they may expect to be compensated from NGOs or from the government programs (Gladfelter 2018, 18).

Climate change will undoubtedly continue to impact water access and distribution in the future. According to scholar Bhaskar Shrinivasulu Padigala,

> Increasing glacial discharges are likely to rise for some time, but the water flow is then expected to reduce with decreasing glacier size. The effects of this phenomenon are most likely to be felt by communities occupying the arid parts of the region and who are heavily dependent on snow melt water for their livelihood.
>
> (Padigala 2017, 9)

The systems for water management in Ladakh must be built to handle both surplus water and extreme scarcity, and to recognize that the future of water access under climate change will shift. The design interventions that support irrigation will therefor need to be flexible, mutable, and resilient.

Note

1 Dawa et al. describe one village *pabchu* situation in which at least 40 different people are needed to guard *yura* for a 36-hour period. One agricultural season will have multiple *pabchu*, so this represents a huge investment of people's time and oversight (Dawa, Dana, and Namgyal 2000, 250).

References

Acharya, Somen, A.K. Katiyar, V.K. Bharti, Guru Charan, B. Prakash, and R.B. Srivastava. 2012. "Assessment of Irrigation Water Quality of Cold Arid Ladakh Region." *Journal of Soil and Water Conservation* 11 (4): 311–15.

Angchok, Dorjey, and Premlata Singh. 2006. "Traditional Irrigation and Water Distribution System in Ladakh." *Indian Journal of Traditional Knowledge* 5 (3): 397–402.

Bell, C. 1928. *The People of Tibet*. Oxford: Clarendon Press.

Crook, John, and Henry Osmaston. 1994. *Himalayan Buddhist Villages: Environment, Resources, Society and Religious Life in Zangskar, Ladakh*. Bristol: University of Bristol.

Daultrey, Sally, and Reuben Gergan. 2011. "Living With Change: Adaptation and Innovation in Ladakh." *Climate Adaptation*. Available: www.yumpu.com/en/document/view/25089047/living-with-change-adaptation-and-innovation-in-our-planet.

Dawa, S., D. Dana, and P. Namgyal. 2000. "Water Harvesting Technologies and Management System in a Micro-Watershed in Ladakh, India." In *Waters of Life-Perspectives of Water Harvesting in the Hindu Kush-Himalayas: Volume II*, edited by S.R. Chalise and M. Banskota, 235–59. Kathmandu: ICIMOD. http://lib.icimod.org/record/22407.

Gladfelter, Sierra. 2018. *Ladakh's Artificial Glaciers, Ice Stupas, and Human-Made Ice Reserves*. Fulbright-Nehru Student Research Report. New Delhi: United States-India Educational Foundation.

Gutschow, Kim. 1993. "Lords of the Fort, Lords of the Earth, and No Lords at All: Politics of Irrigation in Three Tibetan Societies." In *Recent Research on Ladakh 6: Proceedings of the 6th International Colloquium on Ladakh, Leh, 1993*, edited by Henry Osmaston and Tsering Nawang, 105–15. Delhi: Motilal Banarsidass Publishers.

Higgins, A. Kathleen. 2012. *Artificial Glaciers and Ice-Harvesting in Ladakh, India as an Adaptation to a Changing Climate*. New Haven, CT: Yale School of Forestry and Environmental Studies.

Norberg-Hodge, Helena. 1991. *Ancient Futures : Learning from Ladakh*. San Francisco: Sierra Club Books.

Nüsser, Marcus, Susanne Schmidt, and Juliane Dame. 2012. "Irrigation and Development in the Upper Indus Basin: Characteristics and Recent Changes of a Socio-Hydrological System in Central Ladakh, India." *Mountain Research and Development* 32 (1): 51–61. https://doi.org/10.1659/MRD-JOURNAL-D-11-00091.1.

Ostrom, Elinor. 1990. *Governing the Commons: The Evolution of Institutions for Collective Action*. The Political Economy of Institutions and Decisions. Cambridge; New York: Cambridge University Press.

Padigala, Bhaskar Shrinivasulu. 2017. "Traditional Water Management System for Climate Change Adaptation in Mountain Ecosystems." In *Traditional Water Management System for Climate Change Adaptation in Mountain Ecosystems*, 630–56. Hershey, PA: IGI Global.

Rizvi, Janet. 1998. *Ladakh: Crossroads of High Asia*. Delhi; New York: Oxford University Press.

Sharma, Arjun, and Kunal Bharat. 2017. "One's Waste, Another's Right: Translating History and Making the Ladakhi Commons." In *IACS Conference Proceedings*, 1–28. www.iasc2017.org/wp-content/uploads/2017/06/11N_Arjun-Sharma.pdf.

12 Tree planting and tree armor

In Ladakh, trees are vital for the production of building material, as a means of developing agricultural lands, and in helping to form identifiable village space. Yet, in this high-altitude, arctic desert region, forests are essentially nonexistent, and even individual trees need to be supported into maturity. Open grazing practices and the challenging environmental conditions of the region limit overall vegetation growth, and necessitate specific strategies for supporting tree health.

Trees provide critical human and ecosystem services, providing animal habitat and shade for buildings, as well as producing a variety of food, fuel, and building material. In Ladakh, trees also figure into the spatial and environmental practices that impact land management, and in turn occupy a significant position in village cultural landscapes (Figure 12.1). Trees therefore play a number of different roles in Ladakhi space and society, each with associated management strategies.

In the context of climate change, trees will likely experience increasing environmental stress, especially in the desert landscapes of Ladakh. Material and resource scarcity already constrain development here, and under climate change this environmental exposure is only expected to intensify. However, just as the environmental conditions in Ladakh might shift with changing weather patterns, land use, and demographic trends, innovative approaches to planting, management, and protection can be leveraged in defense of trees.

Trees are one example of the host of environmental resources in Ladakh with a particular role to play in climate change adaptation. Moreover, because village groups already practice unconventional tree husbandry techniques and specific strategies for communal management and use, trees might be considered a useful case study in the topic of climate change adaptation. Against the backdrop of a changing climate, the husbandry, cultivation, and creative land management strategies developed in support of regional trees stand out as a critical piece of longer-term sustainability measures. In this context, widespread tree planting efforts, as well as both new and age-old Ladakhi practices for the armoring of trees, offer a unique approach to the management of trees within a landscape commons.

Figure 12.1 Mature trees line the main highway in Ladakh, framing village lands and providing a shaded circulation corridor.

This chapter highlights both new and longstanding tree husbandry practices in Ladakh, and then considers the potential for trees to provide specific benefits in a future climate-adaptive landscape. Current tree husbandry strategies are presented as a taxonomy of different design approaches, to illustrate the design thinking that underscores these practices. In so doing, trees stand out as both a critical component of current land use structures, as well as a critical feature of climate change adaptation in the region.

Background

Ladakh's climate and relative aridity is particularly limiting for tree survival and diversity. The majority of the region's trees can be found at low elevations, along river corridors and at sites of human settlement. However, the cultivation of even a single tree in this landscape typically requires judicious management and careful attention, first in early stages while the tree is gaining a foothold in the soil, and then throughout its life, as grazing herds threaten its existence. Ladakh is not a place where trees naturally grow with ease, and, indeed, forests are simply not a feature of this landscape (Ferrari 2018).

Tree planting underpins the founding of the first Ladakhi villages, as described by the earliest known records of the area. According to Dollfus et al., the act of planting a stick in the ground worked as both a soil fertility test and a way of determining whether it was a good idea to settle in that location. They describe a process in which the budding of a newly planted tree meant that "the decision is taken to establish a village or a fortress in this place" (Dollfus, Lecomte-Tilouine, and Aubriot 2009, 283). These stories come from accounts dating back to 430 bc, when the valleys of north Ladakh were first settled. At that time, Dollfus et al. noted that the trees planted then were typically "walnut, birch, willow or fruit trees" (Dollfus, Lecomte-Tilouine, and Aubriot 2009, 283).

These accounts demonstrate the foundational role of trees in the origin story of Ladakhi culture and civilization, and particularly highlight their use in determining land suitability, as well as their ability to provide food products. Even at the earliest stages of Ladakhi settlement, just as in the present day, the act of site development involved tree planting. Unlike in many other regions of the world, where tree clearing offers a sign of a territory under human development (Rutkow 2013), in Ladakh tree planting marks an intention to control, contain, and civilize space.

Accounts written by European visitors to the region also acknowledge the value of trees in the Ladakhi landscape, particularly in terms of fruit and nut production. In 1631 and 1715, respectively, travelers' journals cite the delicious apples found in the region, as well as the intermittent stands of apricot trees (Dollfus, Lecomte-Tilouine, and Aubriot 2009, 286). These early accounts provided by foreigners substantiate the agricultural records from the area that suggest that trees were intentionally cultivated for food products. Later, in William Moorcroft's early nineteenth century journals, he notes that wood from willow trees was used in Ladakhi tools, such as the sturdy ploughs found across the region. Finally, in records of early Ladakhi construction techniques, there is evidence that vernacular buildings relied upon local trees for framing systems and to support flat earthen roofs (Khan, Bray, and Devers 2014; Ferrari 2018).

Trees in Ladakh

Two major trees typify the Ladakhi landscape: poplar trees (*Poplus sp.*) and willow (*Salix sp.*). These two varieties are especially well suited to the climate, and have been used extensively for timber lattice framing buildings, producing *talu* ceilings, and making furniture and household objects (Figure 12.2). In addition to these two major tree types, the tamarisk (*Myricaria sp.*) and sea buckthorn (*Hippophae rhamnoides*) are found in many locations, while juniper *(Juniperus sp.)* is a rare and venerated tree species. A number of varieties of cultivated fruit trees have also been introduced, such as apricots, walnuts, and apples, and these trees produce food products that are an important part of the Ladakhi diet.

Trees play an important role in Ladakh, from a social, environmental, and physical place-making perspective. They provide wildlife habitat, for birds and other animals. Fuel is provided primarily by dung collected in the open-range landscape, but also can come from woody species, such as *Caragana brevifolia* and *Artemisia spp.* (Dollfus and Labbal 2009). Trees are used in vernacular masonry buildings, forming the roof structure for flat earthen roofs and also critical structural framing components (Khan, Bray, and Devers 2014). Even apricot pulp, coming from local fruit trees, was historically used as an adhesive in the production of plaster (Feiglstorfer 2014, 377). Beyond this physical support, trees serve to provide shade to buildings, limiting overheating from direct solar gain. Finally, trees articulate and characterize human settlement in the area, by defining spaces under cultivation, by producing a village wood lot, or simply in distinguishing village lands from the natural mountain landscapes beyond.

Figure 12.2 Timber lattice structural framing system holds smaller wooden talu ceiling components.

Despite the many uses for trees in Ladakh, and attendant cultural and social appreciation, large stands of trees are still relatively rare in this context. The climate, characterized by cold winters and widespread aridity, makes it difficult for trees to thrive, particularly when they are saplings. Water scarcity is a limiting factor, as trees require regular irrigation across the seasons, especially in their first year of life. A second challenge to tree health comes from the roaming herds of livestock. These animals routinely graze on all types of vegetation in the landscape, impacting the survival rates of young trees. However, the many mature trees in Ladakh do tend to be well suited to the climate and can thrive in the climate and soil, when adequately protected and watered.

Trees are also used, today as well as in the past, as spatial markers at sites of development and in territorial expansion. The land planted with trees, as in a field under intentional cultivation rather than a natural forest, is called a *lcang sa* in Ladakhi. According to Dollfus and Labbal, the word *lcang* can mean both tree in general, and, in particular, willow, while other types of trees might generically be called *shing*, the word for wood (Dollfus and Labbal 2009, 96). According to Dollfus and Labbal, the planting of poplar and willow trees is not only used as a technique by farmers to claim space, but also to make money (Dollfus and Labbal 2009, 100). Payments provided by the Indian government range from two to five Indian rupees per planting, and can also finance the protective fencing surrounding fields. In villages where families increasingly lack the manpower needed for farming efforts, tree plantations provide a relatively low-maintenance, high-value crop option.

The Indian government has encouraged tree planting in Ladakh in part because trees help to sustain a local economy, when they are used for material. Locally available wood products reduce the need for imported goods from other areas, with their high transportation costs and a seasonally impassable road network. Wood, fruit, and nut harvesting marks just one of the ways in which the region sustains its autonomy and independence.

Social and cultural values also provide powerful reasons for tree production in Ladakh. Historically, trees have been highly prized by Ladakhi people, primarily for the material resources and food that they provide. Over time, these benefits have helped to build the perceptions of abundance, reverence and social status that Ladakhis confer with trees in the landscape. Trees are considered by many, for instance, to be extraordinarily valuable, and a sign of wealth and abundance in an otherwise austere environment. Trees found in natural areas, such as the single standing trees located on rocky mountain surfaces, can be considered occupied by deities (Ferrari 2018). It makes sense that in the largely treeless and profoundly arid landscapes of Ladakh, trees would be acknowledged as an exceptional feature. While they do bring real value in terms of building supplies, fuel or protection from the sun, beyond these concrete benefits, trees occupy an esteemed position that transcends the functional.

Husbandry

In many contexts, the various support systems that shelter, protect, and sustain trees rarely find prominent display. Instead, the irrigation systems, soil composition, and the symbiotic relationships between animals, humans, and trees remain largely invisible, or structured in such a way that they appear to be knit into the fabric of Ladakhi villages. And yet, many design interventions exist to support trees in this context, including physical separation strategies, elaborate watering systems, and social or cultural expectations to promote protection and preservation.

Tree planting done by individuals on their own land for their own gain includes cultivation for personal use, as a means of claiming or developing land, or as a strategy for making money and material to sell. In villages where woodlots are held in common, trees are planted for communal use, and husbandry of that resource is also shared. Use of land held in the commons for tree production allows an entire space to be irrigated using shared water rights, for instance. In addition to these two zones, the private household land holding and a shared tree lot commons, many trees have been planted along public pathways in a less structured but inherently civic manner.

The type and quality of tree protection devices, and various approaches to tree husbandry, varies depending on the type of tree and associated location in the context of Ladakh. On private lands, farmers and households may spend a great deal of time and resources to isolate, grow, and then protect trees. These trees could be apricot trees, shade trees growing near a house, or a field of saplings planted as an investment for wood consumption.

At the village scale, trees might form a common woodlot, or greenspace. In these areas trees signify communal value as well, providing a measure of wealth for the village group. Trees from these lots can be harvested and sold to other villages, used for the needs of that village, or simply saved to create a village landscape.

Trees that exist in the vast stretches of land between villages in Ladakh may support shepherds or nomadic groups with temporary refuge. They may mark a camp or a site of special significance. Some are considered to be sacred, and these trees are protected by cultural values that prohibit them from being cut down (Ferrari 2018). Many of these trees appear to have benefitted from various human interventions, although it is less clear from who, and when these measures were taken. This type of altruism supports the theory that trees hold significance in Ladakhi landscapes, regardless of ownership.

Tree armor

In Ladakh, trees must be protected by fencing or with armor from range animals. Herds of yaks, *dzos*, sheep, cows, and goats move freely through the landscapes outside of villages, where they graze on vegetation, including trees. Tree armor in Ladakh varies in terms of design strategy, material, and lifespan. Almost all of the armoring strategies that are commonplace use some sort of physical barrier method, and many of these interventions are unabashedly on display.

Trees planted in fields, around houses, and even along village footpaths often use an extensive physical enclosure. These systems include wooden fences or metal screens, but most often take the form of a simple masonry wall (Figure 12.3). The masonry wall is, in Ladakh, one of the most fundamental building elements, and can be made of stone sourced from on-site, from rock brought in from elsewhere, or site-produced mud bricks. These walls are ubiquitous features in the desert landscapes of Ladakh, often built to form physical enclosures to isolate various fields or household space.

Another enclosure technique that is commonplace in tree protection is the isolated barrier method. Here, a single barrier is created around the tree trunk, as a form of armor. This approach can employ any type of natural or artificial material, and tends to be implemented in an ad hoc way. It is not unusual to see stacked rocks encircling trees out in the mountains, and in villages and towns, cans may be stacked up along a single trunk (Figures 12.4 and 12.5). Plastic bottles, and other round vessels that would otherwise need to be buried as trash, find new use in encircling tree trunks and limbs. Fabric, or burlap, or other sheet goods

Figure 12.3 A rock wall serves as a barrier in Ladakh.

Figure 12.4 A stone wall built around a newly planted tree.

Figure 12.5 Repurposed cans have been threaded around saplings to protect the trunks.

are also used as armor, wrapped around the base of saplings to deter grazing animals.

For the farmers or villagers who have a vested interest in helping a certain tree thrive, armoring strategies make a great deal of sense and are worth the additional investment of time, labor, and material. However, tree armor is also found across the mountain grazing landscapes, in space that is not owned by a particular village or individual. In these cases, the armoring of trees appears to be done by individuals who have a broader conservation ethic, or interest in promoting trees for their own inherent value.

This approach, which enables cultural and religious interests to shore up environmental stewardship in Ladakh, is a powerful conservation technique. Scholar Susan Darlington outlines a similar strategy for tree preservation in Thailand, in her book called *The Ordination of a Tree* (Darlington 2012). In this context, Buddhist monks have ordained trees in order to assure their survival, particularly in the face of imminent threats. While the sacred qualities of trees in Ladakh tend to lack prescriptive or ceremonial religious intervention, they still occupy venerated physical space in mountain landscapes.

Conclusion

One of the major challenges to development in Ladakh is material scarcity. Under climate change predictions, resource challenges are only expected to grow. This backdrop not only necessitates judicious resource management methods, but also suggests that design ideas that could support existing resources, or even foster the development of new stocks, is critical.

As a carbon sink, trees help to mitigate climate change, and can provide useful shade, habitat, and local materials that might broadly benefit people under climate change stresses. Trees are a protected part of the landscapes of Ladakh, for both their physical and sacred offerings. But beyond these explicit uses, trees may also help villages to transition away from traditional farming products to more extensive fruit production, subsidized tree production, or perhaps even tourist attractions.

In signaling a commitment to safeguard urban tree canopy, these design interventions could create a new brand of signage to spur engagement and awareness both in Ladakh and beyond. Trees, after all, orchestrate a host of co-benefits; they provide animal habitat, produce food and material products, they create micro-climates and even fix the air. But the average lifespan of a planted tree in contemporary urban areas is well under ten years, and impending climate change only threatens to exacerbate existing environmental stressors (Marritz 2012). Looking beyond Ladakh, the measures taken to armor and support trees in a challenging desert environment stand out as an affordable, scalable and practical stewardship model.

Bibliography

Darlington, Susan. 2012. *The Ordination of a Tree*. Albany: SUNY Press. www.sunypress.edu/p-5586-the-ordination-of-a-tree.aspx.

Dollfus, Pascale, and Valérie Labbal. 2009. "Ladakhi Landscape Units." In *Himalayan Landscapes Over Time: Environmental Perception Knowledge and Practice in Nepal and Ladakh*, edited by Joëlle Smadja, 85–106. Pondicherry: Institut Francais de Pondichéry. www.abebooks.com/Reading-Himalayan-Landscapes-Over-Time-Environmental/1386152000/bd.

Dollfus, Pascale, Marie Lecomte-Tilouine, and Olivia Aubriot. 2009. "Agriculture in the Himalayas: A Historical Sketch." In *Himalayan Landscapes Over Time: Environmental Perception Knowledge and Practice in Nepal and Ladakh*, edited by Joëlle Smadja, 280–323. Pondicherry: Institut Francais de Pondichéry.

Feiglstorfer, Hubert. 2014. "Revealing Traditions in Earthen Architecture: Analysis of Earthen Building Material and Traditional Constructions in the Western Himalayas." In *Art and Architecture in Ladakh*, edited by Erberto F. Lo Bue and John Bray, 364–87. Boston: Brill.

Ferrari, Edoardo Paolo. 2018. *High Altitude Houses: Vernacular Architecture of Ladakh*. Florence, Italy: Didapress.

Khan, Kacho Mumtaz Ali, John Bray, and Quentin Devers. 2014. "Chigtan Castle and Mosque: A Preliminary Historical and Architectural Analysis." In *Art and Architecture of Ladakh*, edited by Erberto F. Lo Bue and John Bray, 254–73. Boston: Brill.

Marritz, Leda. 2012. "A Million Trees? Only If We Can Keep Them Around." *Next City*, April 18, 2012. https://nextcity.org/daily/entry/a-million-trees-only-if-we-can-keep-them-around.

Rutkow, Eric. 2013. *American Canopy: Trees, Forests, and the Making of a Nation*. New York: Scribner.

13 Food security

During the course of the past five decades, Ladakh has become better connected than ever, as an increasing network of roads, airline routes, and helicopter service have helped to open the region up to the outside world. Faster communication methods, including limited internet access, cell phone service, and even Amazon deliveries have radically improved connectivity. Finally, throughout the post-colonial years the region has been well supported by a protective state, which offers aid and employment that strengthens bonds between Ladakhi people and the Indian government. Together these factors paint a picture of a region that may be physically removed from the rest of the Indian subcontinent, but is by no means entirely isolated. Moreover, as interaction with other places grows, the relative autonomy of Ladakhi society appears to decline.

Ladakh's newfound accessibility and growing interdependence brings about benefits as well as challenges, particularly for regional food access. The increased physical connections, communication outlets, and state assistance enable communities to import food. As traditional agricultural practices shift to accommodate more advantageous trade opportunities, village communities will experience an erosion in overall food production and concomitant self-sufficiency. These demographic and lifestyle factors undermine an otherwise stable foundation for food security in the region, upon which climate change is an additional stressor.

While new modes of connectivity may improve the population's access to food and assistance, they do little to mitigate the specific weather and geographical factors that influence food sovereignty in Ladakh. Indeed, older methods used for food provisioning, such as those that demonstrate a reliance on local systems and products, may be most useful tools for managing the challenges of climate change today. Climate change is inextricably linked to food, and therefore survival. Author Jared Diamond notes the broad importance of food security in his book on population collapse, especially with regard to regional connectivity. He notes that "Compounding these problems of climate change, many past societies didn't have "disaster relief" mechanisms to import food surpluses from other areas with a different climate into areas developing food shortages" (Diamond 2005, 12).

Despite all of the visible areas of growing connectivity in the region, Ladakh remains one of those places that is still relatively poorly connected to other areas due to geography and weather, and so food security under climate change then also becomes more tenuous.

The traditional diet

Ladakhi agriculture has historically been finely tuned to the trans-Himalayan context, limiting any need for external inputs and well adapted to the carrying capacity of the land. Staple crops grown in this context formed the foundation of the traditional diet, and satisfied the fundamental nutritional needs of Ladakhi people and their livestock. According to scholar Vladimiro Pellicardi, "In the past, food-grain security was assured by local production of barley, wheat and, in small amounts, other millets" (Pelliciardi 2013, 113). Additional vegetables, such as peas, and fruits, as well as oils and animal products helped to round out the traditional diet.

Although Ladakhi settlements appeared to be relatively self-sufficient and food secure over the years, area population was effectively curbed by the productive capacity of the land (Crook and Osmaston 1994; Norberg-Hodge 1991). Dollfus et al. note that "Ladakhi farmers, who did not practise fallow and crop rotation, but obtained an average of ten times the quantity of seed sown (and even twenty times in Dras, in the west of the country" (Dollfus, Lecomte-Tilouine, and Aubriot 2009, 289). Their chapter chronicles the substantial yields achieved by Ladakhis even in the 1800s, when travelers such as Moorcroft and Trebeck first ventured into the area.

However, despite this relative stability, there is evidence that the traditional Ladakhi diet also lacked critical vitamins and minerals. For instance, William Moorcroft's 1818–1819 journals note the prevalence of gout (an ailment caused by iodine deficiencies) in a number of the villages he visited, as well as his own difficulty in transporting food through even the well-worn trade passages of Ladakh's high-altitude mountains (Moorcroft 1841). Throughout the nineteenth century, the limited diet available to Ladakhi villagers lacked nutritional diversity, and the relative isolation of the area required communities to satisfy the vast majority of their own dietary needs.

Food insecurity

Despite the challenging environmental conditions, and the difficulty of growing diverse food products, Ladakh appears to be a good place to grow food. According to Pascale Dollfus, "Unlike the Kathmandu Valley in Nepal, until the introduction of maize in the 17th century and potato in the 18th century to complement rice, Ladakh does not seem to have suffered any serious famine" (Dollfus and Labbal 2009). Henry Osmaston cites high yields despite the short growing season for staple crops in Zanskar in the

1980s, which he attributed to a combination of judicious resource management and the use of excellent regionally adapted crop strains (Osmaston 1994).

Even so, in the past Ladakh remained exposed to challenging environmental factors, which could more or less be planned for, and a series of more irregular threats. Scholar Quintin Devers suggests that the earliest periods of the Ladakhi state were characterized by banditry and political instability (Devers 2018). William Moorcroft's journals make note of raids as well, including events that he witnessed during his travels in the early nineteenth century (Moorcroft 1841). According to Moorcroft, a raid on a village could result in the abrupt loss of grain stores, herd animals, and valuable goods. This kind of instability was not just vexing for villagers but also dangerous, as putting up adequate stores of food to outlast the long winter months was, for individual farmers, critical to survival.

The result of challenging environmental and social conditions for food security in Ladakh gave rise to a series of finely tuned and well-tested habits for growing, sharing, storing and caching foods. Over time, Ladakhi villages became highly self-sufficient if not entirely food secure (Dollfus, Lecomte-Tilouine, and Aubriot 2009). The relative self-sufficiency seen in Ladakh in the mid-twentieth century stemmed from highly tuned agricultural methods and self-help strategies for keeping provisions from being stolen or damaged. Food security drew upon this relative autonomy but then layered over additional support provided by different groups, each of which acted as a source for collective food aid.

Even today, food security remains strong in Ladakh: here communities may experience some hunger but not famine (Dame and Nüsser 2011). Visiting non-Ladakhi laborers and nomadic herders may be the region's most vulnerable individuals to food stress. The strong food self-sufficiency of most residents has enabled the region to act stable, even though the relative food security of the overall system is limited by tenuous environmental and social conditions.

Traditional food aid practices

Historically, Ladakhi monastic leaders held an important role in providing a safety net for area food security. Each monastery maintained large stores of grain that served a primary purpose of feeding their own residents and a secondary service of filling provisioning gaps in the region. These grain reserves were built up by a village-structured tax on cropland; each farming family was responsible for delivering a specific portion of their yield to the monastery as a means of tithing. Such donations to the monastery were absolutely codified, which enabled the monastery to amass a significant amount of material wealth. In addition to monthly food offerings, monks received payment for services rendered, so that religious events, such as funeral services, provided additional opportunities for the generation of income.[1]

A large reserve of food products, such as grains that could be stored in containers for decades, could then serve as a food bank for multiple villages overseen by a single monastery. The practice of *shushok* allowed needy farmers to borrow food directly from the monasteries, and then repay that debt with interest over subsequent years. According to researchers Juliane Dame and Marcus Nüsser, "In case of harvest failure, staples had to be borrowed from monasteries or well-off households and returned with high interest rates" (Dame and Nüsser 2011, 185). Such usury could devastate farmers operating households within meager production margins. However, *shushok* also provided an important safety net, as monasteries could immediately intervene with helpful resources in the case of wide-spread crop failure, political unrest, or other major losses.

Beyond monasteries, another important social support system at the regional scale came in the form of inter-village alliances. Ladakhi villages have historically worked together and helped each other in times of need, and many continue to maintain these bonds today. For instance, long-running inter-village relationships produce marriages and other forms of reciprocal support, such as food, material trading, or physical security.

Finally, at the scale of the Ladakhi household, cooperative groupings within a single village also enabled fundamental assistance. Individual families aligned themselves with other households within a single village, creating smaller synergistic groupings for mutual aid. These alliances allowed farming households to share tasks that would otherwise be overwhelming for one family to manage alone, such as specific farming work, food processing, or large construction efforts.

Ladakhi social, cultural and religious customs thus were tailored aid to various needs at different scales. While monasteries may serve multiple villages in an area, and provide material resources at a cost, they could intervene in cases where entire crops failed, or whole villages experienced a crisis. Inter-village relationships enabled practical exchanges of goods and individuals that took advantage of qualities provided by physical distance. At the scale of the village, household groupings created mutual aid to cover gaps in labor, materials or provisions. Together, these three different levels of religious and social assistance have historically worked in parallel to provide a scaled insurance system for the whole region.

Current practices

Today, these three levels of assistance (the household group, the village, and the monastery) remain mostly intact, especially in small rural villages. In most areas, monasteries still receive regular monthly donations from Ladakhi farming households, but the use of *shushok* has dwindled (Figure 13.1). A fourth, larger source of support provided by the Indian state has infiltrated all scales of villages in Ladakh through the ration program. And, finally, NGOs and various visiting benefactors have intervened in

Figure 13.1 A monastic storeroom today.

individual cases to provide food and material support. These five support systems work side by side in Ladakh to ensure relative food security in the region.

Food production in Ladakhi villages remains fairly constant, but the nature of the work is changing. The population grows, largely through national and international visitors, seasonal laborers and all of the entrepreneurs that come into the area for employment and other types of opportunity. More people are now moving to regional cities and towns, and children who once provided critical workforce on farms may instead choose to attend regional boarding schools for secondary and post-secondary education.

The types of food produced on farms in Ladakh is also changing. Increasingly farmers augment staple crops and animals with vegetables and other market crops. Fewer households are keeping animals, due in part to the difficulty of maintaining adequate food for livestock, and in part because available farm labor is diminishing under new demographic trends. Despite these adjustments to agriculture, staple crops remain strong in Ladakh. According to scholar Vladimiro Pellicardi,

> The total area under barley and wheat in Leh District has remained almost constant during 2001–2008 (around 7,400 hectares); the fields under barley, decreased of 282 ha (–6%), is stabilized since 2004 around 4,452 ha, while fields under wheat, increased of 364 ha (+14%), is stabilized around 2,968 ha (SHB 2009).
>
> (Pelliciardi 2013)

In traditional Ladakhi households, food storage spaces continue to occupy a considerable building footprint (Ferrari 2018). These stores are often located on the northern or western side of the building, and can be buried outside or located at the ground floor level, where they are easy to access during the course of the day and year, but also positioned to take advantage of natural cooling (Figure 13.2).

Food storage with refrigeration remains extremely rare in Ladakh, due primarily to the limited availability of electricity for freezers and refrigerators and the frequent brown-outs that occur even in the large town of Leh. While generators are used to fill this gap, many individual households prefer to rely upon more traditional, low-tech food storage strategies. These include grain storage in large cool interior rooms, and using rooftops to dry vegetables and meats for use in the future. Outside of the building, Ladakhis use pits dug out of the ground to keep root vegetables cool and protected. These pits are then covered over with earth or a heavy lid to hold produce over many months.

The foods that can be stored underground include potatoes, grains, onions, pulses, roots. In pantries, dried fruits, herbs, and vegetables are stored in containers. Nuts harvested from apricot seeds are common, as well as dried meat. Households with cows or *dzos* keep fresh milk available through the winter. However, eggs are rarely produced on site in villages, as chickens are too difficult to keep alive during the cold months. While storerooms tend to have physical vessels that can be removed, the oldest Ladakhi houses have grain storage containers that are literally built

Figure 13.2 The hole in the foreground of this photograph is used to store root vegetables outside.

into the masonry structures, spaces protected from pests as well as thermal fluctuations.

Current ration

Food quantity and quality has improved in Ladakh in the past several decades, as more diverse food products have entered the market through trade and the government ration (PDS). The highly subsidized PDS allowance, which consists of sugar, oil, flour and rice, undoubtedly contributes calories to the Ladakhi diet (Figures 13.3 and 13.4). Unfortunately, these products come with a nutritional deficiency that may detract from overall health and self-sufficiency. According to the Ladakh Autonomous Hill Development Council (LAHDC), "Ladakh is getting excessively reliant on the outside world for critical needs such as food."[2] In terms of food security, both the availability and reliability of this ration is problematic.

Food dependence has an unusually high cost for Ladakh. Because the region is isolated from other parts of India due to road closures during the winter months, and lacks trade access with the bordering nations of China and Pakistan, the region can effectively become a trade island. Imported

Figure 13.3 A ration truck delivers foodstuffs to Ladakh via road.

Figure 13.4 A ration storeroom and dispensary, outside of Leh.

food can become excessively costly here due to long road mileage and limited year-round availability. Yet, as farmers have transitioned into other lines of work, food security and independence no longer can be assured at the level of the household. While many families in Ladakh (particularly in rural villages) still rely on subsistence agricultural practices, the Indian ration system and relatively open trade has helped to boost reliance on other food provisioning systems.

Scholar Vladimiro Pellicardi notes that grain imports have grown significantly in the twenty-first century, where:

> The annual deficit of food-grain, as a difference between the quantity required and the quantity available (locally produced), is imported via commercial traders, cooperatives and the Consumer Affair Department (CAD). During the past 10 years, the quantity of food-grain imported in this District by PDS has increased from about 56000 quintals, in 2000/01, up to 103000 quintals in 2009/10 (61000 rice + 42000 wheat flour called "atta"), catering for about 111800 souls through the 130 sale outlets.
>
> (Pelliciardi 2013, 111)

Scholars suggest that the subsidies and handouts provided by the ration system have served to undercut the self-sufficiency that once characterized Ladakh. According to scholar Sonam Dawa, "This has virtually destroyed such qualities as self-reliance, sustainability and even self-respect; so vital for an area that remains cut-off from other parts of India and the world for many months every year" (Dawa 1999, 375–6). According to Dawa, the outsourcing of food products leaves Ladakhis vulnerable to the weather, where "the late opening or early closure of the road activates a state of alarm in the minds and offices of those responsible for procurement, transport, stocking and distribution of essential commodities" (Dawa 1999, 376). The precariousness of this situation could be exacerbated under the challenges of climate change, where additional environmental stressors impact food dispersion from both local and national sources.

Conclusion

Ladakh has witnessed significant change during the last four decades, with the rise of tourism and an increase in army service and government employment. Scholar Mohammad Deen Darokhan cites a concomitant deterioration in the practice of ancient farming techniques and agricultural reliance. In the past, because of the remote and autonomous nature of the district, Ladakhi people had to be self-sufficient. Today, the strong legacy of self-sufficiency and food security in Ladakh has become threatened by current lifestyle and demographic trends.

It is in the context of transformation and re-imagining under climate change that Ladakhis can correct areas of vulnerability and possibly also find additional benefits. Sites of adaptation include new forms of agricultural entrepreneurship, through the development of large-scale local food provisioning, for sale to the government, the Indian army, tourists and locals.

One major area for adaptive design thinking might be found in simply returning to the age-old practices of food storage, which would enable communities to be self-reliant in the event of a political or environmental crisis. After all, the systems are already in place in Ladakh, with storage practices that have been designed to work with the weather and the crops, with social and religious relationships that reflect established values and altruism, and with a diet that is relatively stable and healthy. A return to traditional storage practices could provide a foundation of food security without impeding other market and social opportunities; it could be a relatively small adjustment that could be the insurance against current vulnerabilities.

Moreover, a return to well-tuned age-old practices does not necessarily need to imply backsliding or changing course. Instead, it is a moment in which the best ideas from the past might be overlaid on the new needs and opportunities of the present. Ladakh's current food system already knits together old agricultural strategies and new commodities. If this provisioning patchwork is envisioned with an aim toward affirming food security, it could be bolstered even further. This could be an important model for other areas too; as climate change and associated food insecurity begins to be felt in other parts of India and across the globe, Ladakh has an important model for self-sufficiency to share.

Notes

1 Interestingly, nunneries received far fewer material benefits than monasteries, and lack the ability to earn payment (Gutschow 2004).
2 LAHDC-L 2005, 2.

References

Crook, John, and Henry Osmaston. 1994. *Himalayan Buddhist Villages: Environment, Resources, Society and Religious Life in Zangskar, Ladakh*. Bristol: University of Bristol.
Dame, Juliane, and Marcus Nüsser. 2011. "Food Security in High Mountain Regions: Agricultural Production and the Impact of Food Subsidies in Ladakh, Northern India." *Food Security* 3 (2): 179–94. https://doi.org/10.1007/s12571-011-0127-2.
Dawa, Sonam. 1999. "Economic Development of Ladakh: Need for a New Strategy." In *Ladakh: Culture, History, and Development between Himalaya and Karakoram*, edited by Martijn van Beek, Kristoffer Brix Bertelsen, and Poul Pedersen, 369–78. Denmark: Aarhus University Press.

Devers, Quentin. 2018. "Archaeological Ladakh: Recent Discoveries Redefining the History of a Key Region between the Pamirs and the Himalayas." *Central Asiatic Journal* 61 (1): 103–32.

Diamond, Jared. 2005. *Collapse: How Societies Choose to Fail or Succeed.* New York: Penguin Books.

Dollfus, Pascale, and Valérie Labbal. 2009. "Ladakhi Landscape Units." In *Himalayan Landscapes Over Time: Environmental Perception Knowledge and Practice in Nepal and Ladakh*, edited by Joëlle Smadja, 85–106. Pondicherry: Institut Francais de Pondichéry. www.abebooks.com/Reading-Himalayan-Landscapes-Over-Time-Environmental/1386152000/bd.

Dollfus, Pascale, Marie Lecomte-Tilouine, and Olivia Aubriot. 2009. "Agriculture in the Himalayas: A Historical Sketch." In *Himalayan Landscapes Over Time: Environmental Perception Knowledge and Practice in Nepal and Ladakh*, edited by Joëlle Smadja, 280–323. Pondicherry: Institut Francais de Pondichéry.

Ferrari, Edoardo Paolo. 2018. *High Altitude Houses: Vernacular Architecture of Ladakh*. Florence, Italy: Didapress.

Gutschow, Kim. 2004. *Being a Buddhist Nun: The Struggle for Enlightenment in the Himalayas*. Cambridge, MA: Harvard University Press.

Moorcroft, William. 1841. *Travels in the Himalayan Provinces of Hindustan and the Panjab; in Ladakh and Kashmir; in Peshawar, Kabul, Kunduz, and Bokhara.* Vol. 2. London: John Murray. www.wdl.org/en/item/17537/.

Norberg-Hodge, Helena. 1991. *Ancient Futures: Learning from Ladakh*. San Francisco: Sierra Club Books.

Osmaston, Henry. 1994. "The Farming System." In *Himalayan Buddhist Villages: Environment, Resources, Society and Religious Life in Zangskar, Ladakh*, edited by John H. Crook and Henry Osmaston, 139–98. Bristol: University of Bristol.

Pelliciardi, Vladimiro. 2013. "From Self-Sufficiency to Dependence on Imported Food-Grain in Leh District (Ladakh, Indian Trans-Himalaya)." *European Journal of Sustainable Development* 2 (3): 109–22. https://doi.org/10.14207/ejsd.2013.v2n3p109.

14 Recruiting allies

Introduction

Over the centuries, Ladakh has been exposed to a wide variety of influences due to its location along major international trade routes. These routes acted as conduits for a steady stream of products, ideas, and people that in turn challenged, stretched, and influenced local norms. Evidence of trade relationships can be charted across the centuries in the region's richly varied architecture, landscape and planning projects (Devers 2017; Lo Bue and Bray 2014). Interaction with outsiders has ebbed and flowed over the centuries, but today the existence of external actors in the region can be viewed as a potential source of support for the development of climate-adaptive design work.

As Ladakhis look for new partners to help produce development projects, there is an opportunity to tap into existing networks of allies already connected to Ladakh through trade, tourism or other types of cultural exchange. A number of significant projects have been realized in collaboration with outsiders in recent years, and the practice has emerged as a strategy in the region's portfolio of climate change adaptation projects. This chapter outlines the types of relationships that have already been established with Indian and foreign allies, and considers future collaborative directions under the development of climate-adaptive design efforts.

Background

In the mountain landscapes of north India, foreign-sponsored design and construction projects present exciting opportunities to enhance, improve, and modernize traditional village life. From an architecture and planning perspective, these projects can serve as the critical first wave of development in a remote area, providing village communities with much-needed schools, clinics, or libraries. This altruistic development ostensibly serves as a token of cross-cultural exchange and goodwill, and volunteer design teams often cite the gratification that comes from working for a "good cause" (Brillembourg, Klumpner, and Coulombel 2011).

On the other hand, scholar Lewis Hyde cautions that even the most well-intentioned gift-giving can harbor disturbing undercurrents, where the "bonds" of gift exchange limit a person or group's sense of freedom, mobility, and autonomy (Hyde 1979, 43). Gifts in the form of built constructs can be especially fraught; physical space inherently impacts social norms and behaviors, and can be difficult to re-gift, remove, or return.

Yet, Ladakhis actively seek out external partners for development projects, making potential donors and collaborators aware of village or household needs and associated charity opportunities. And many of these allies want to leave a positive legacy of support in Ladakh, or contribute in some way to rectifying perceived problems. While the desire to cultivate collaborative projects appears to be mutual, the conditions under which these relationships unfold could be better managed for more equitable outcomes. For Ladakhi stakeholders, the strategy of recruiting allies, and educating them so that they can become valuable collaborators, could become a powerful means of affecting impactful change.

After all, design vision comes from local as well as from outside sources, and development projects will reflect the views of those enlisted to participate. In addition to the growing number of mainstream communication outlets available to Ladakhis, the region is now exposed to ever more design input, both from popular culture venues (through newspaper, film, the internet, and television) as well as from visiting allies. The new ideas, values and standards that have been absorbed by the Ladakhi people in recent years are a testament to the diversity, open-mindedness, and fluidity of a culture that will not be fixed in time or space (Aggarwal 1997). Ladakh is a region in transition, and its evolution reflects a hybridity of design values, needs and aspirations.

Public-interest design

For the designer working abroad, public-interest design projects can provide a meaningful outlet for professional skills beyond market-oriented practice. The translation of this expertise to foreign aid contexts, however, requires a professional grounding that goes beyond importing techniques, technologies, and tools into a new environment. When designers undertake a building commission in their home country, the client–expert relationship is well understood. However, when designers offer pro-bono services abroad, the work can easily become obfuscated by the additional factors that characterize aid in foreign countries. These factors vary greatly depending on the design team and their project, but include considerations such as underlying intentions, levels of investment, power dynamics, and communication challenges. To manage this contextual dissonance, designers who aspire to participate in service work need to cultivate additional competencies beyond those they might use in everyday practice. As a rule, skillful international aid work demands sensitive cross-cultural collaboration and a willingness to work within a host country's existing social, environmental, economic and political frameworks (Wickersham 2014).

As "do-good-design" becomes increasingly understood as complicated, or worse, associated with "architectural imperialism" (Nussbaum 2010), it is doubly important to disentangle and interpret the manifold layers of this form of practice. Meanwhile, in Ladakh, many of the public-interest design projects sponsored by outsiders have begun to establish a highly visible new vernacular architecture (Prakash 1991). The inherent challenges of this type of design production extend beyond trained design professionals to the additional allies who help to facilitate work, including funding groups, individuals, volunteers, and organizational staff.

The discipline of architecture may inherently possess a public-interest bias. It is a profession that concerns itself with the physical provision of basic human needs, such as sanitation and shelter; ethical standards, such as public safety and health; and the improvement of the physical environment, raising overall quality of life indicators and urban functioning. One of the profession's central themes – that design can be used as a tool to uplift the human condition – has been used to describe and brand many different social service projects. This narrative was expressed during the Modernism movement, when practitioners championed the connection between design and service in the name of social progress. In 1953, Pietro Belluschi argued that "architecture is space and form serving a social purpose beyond esthetic satisfaction" (Kahn, Weiss, and Scully 1953, 47), and that era was characterized by schemes to modernize outdated buildings, as well as massive urban upgrading efforts and visionary utopian projects (Pyla 2015).

While this interest in social service has not waned per se, the subject and approach of design aid efforts have shifted in the intervening years. Designers are now far more likely to cite an interest in critical regionalism, post-disaster redevelopment or humanitarian aid (Charlesworth 2014; Brillembourg, Klumpner, and Coulombel 2011). Whereas modernist architects once folded many different so-called noble social goals into their project programs, current practitioners instead tend to identify this work as an independent genre within the profession, requiring unique training, motives, and expertise (Coulombel 2011). Moreover, public-interest design, especially as conceived as a form of international service, has recently moved to the forefront of the design discipline's consciousness; it has become a new mode of practice (Wakeford and Bell 2008).

One of the major challenges for appropriate public-interest practice in international contexts is that designers must learn how to engage in entirely new social, cultural, environmental, economic, and political landscapes. While designers may bring valuable knowledge and skills from their host country, they must translate both their approach and their output to meet the needs of clients living in another region. This is particularly difficult for visitors who spend a limited amount of time at a distant project site, or who were trained in a radically divergent context. Architectural educator Jay Wickersham notes that "Architects working abroad face a further challenge, perhaps even more complicated and confusing: making sense of the social and political projects in which they are involved" (Wickersham 2014). Indeed, beyond the physical building concerns that public-interest designers must address in a foreign setting (such as form, function, environment and aesthetics), they must also understand the sociocultural and political conditions that frame such programs.

Ladakh's neo-vernacular movement

It is in the role of a guest, then, that external allies must operate. As the new community projects sponsored primarily by outside individuals or groups spring up across the region, traditional ideas and construction methods have given way to an equally legible neo-vernacular approach. This design response rejects the idea of replicating traditional buildings, but also refuses to appropriate the contemporary design standards that characterize much of the region's newer market-driven development.

This neo-vernacular style cannot be qualified so much in terms of form or style, but instead due to each building's intentional connection to landscape, reliance on locally sourced materials and adapted construction techniques, and the emulation of functions found in many historic buildings.[1] These traditional impulses are then joined with contemporary global standards for energy efficiency, seismic safety, available technology, and thermal comfort. Thus Ladakh's neo-vernacular echoes the ideals expressed by Kenneth Frampton's "Prospects for Critical Regionalism," sharing some of same physical approaches that appear in traditional buildings, such as similar responses to landscape, climate, and tectonic form (Frampton 1983). Moreover, much of this work moves beyond Frampton's call for a physical response to critical regionalism by also incorporating design thinking that acknowledges the region's distinct social, cultural and political backdrop.

Types of allies working in Ladakh

There are several clear categories of types of allies that collaborate on projects in Ladakh. Many projects have been spearheaded by a visiting designer, or individual, who has a unique connection to the region and a specific vision for a design intervention. Larger, long-term projects tend to come from NGO or international aid organizations, with their considerable experience in development work and mission-driven interests. Government groups provide support services in a variety of sectors. Finally, projects may arise out of research collaborations, or from materials and workshops disseminated by scholars on the area.

While Ladakh has received a significant wave of outside design guidance since the mid-1970s, design aid projects are still sparsely scattered across the entire region. In part due to the late opening of the region to tourism, and in part because the remote and rugged landscape limits access, sponsored work here has been constrained. It could be argued that this relative isolation redoubles the need for design assistance from the state, NGOs and private individuals. Many of Ladakh's villages are well behind other parts of India in terms of access to goods and services: Some villages lack electricity, roads or plumbed water, and across the board schools and clinics tend to be underfunded. Many villages have one or two public buildings, such as schools, community centers, or health clinics that have been sponsored by foreign individuals, NGOs or other aid groups.

The dozens of public-service projects built using outsider sponsorship in the region suggest a fundamental shift away from traditional priorities. Each of these projects offers robust programming for public use, setting new and hopeful directions for the future of this region. In different ways, the projects offer community-oriented buildings, investigating new systems to support, bolster, and sustain traditional Ladakhi culture. However, this programming often reflects outside interests and agendas. For instance, many of the structures house new types of programs not represented by historic buildings, these are Ladakh's new museums, health and dental clinics, and schools.

Individual sponsorship

Ladakh has long been a destination for foreign trekkers, travelers, and scholars. More recently, the numbers of Indian tourists have swelled in the region as well. These individuals and groups create lasting bonds with the Ladakhi people that they meet, and can return many times over the course of a lifetime. As a result, Ladakhis have cultivated relationships with donors for decades. Early efforts were characterized by educational sponsorship for Ladakhi youth, but that foreign aid has, over time, extended beyond tuition support to influence the built environment as well. Today it is not uncommon to see projects in Ladakh that were spearheaded by individuals or groups who bring their own values about healthcare, education, or cultural preservation.

Occasionally individuals collaborate with their own external design teams, or groups of volunteers, in order to coordinate large-scale development efforts over time. In other cases support and sponsorship is more ad hoc, involving the gifts of funds to support Ladakhis in their own design visioning, or with gifts of materials, technologies, or labor. These methods for supporting development in Ladakh lack a single organizational framework, but have effectively ushered in change over time.

One example of this types of engagement for climate-adaptive projects include a series of artificial glaciers commissioned above the village of Stongde, in Zanskar, by the American educational group called *The Institute for Village Studies*. Students raised money, coordinated communication between the lead Ladakhi designer and the village stakeholders, and then left oversight of the construction process to local stakeholders. Several villages in the region have solar buildings built by individuals and organizations as well. One school in Zangla incorporates passive solar orientation, a trombe wall, and enhanced insulation. This building was designed by

Figure 14.1 A flyer advertises volunteer opportunities to work on the school in Zangla.

Figure 14.2 Csoma's Room Foundation worked to manage the design and build of this small school building in Zangla.

Figure 14.3 The façade of the school building opens to the south and incorporates a trombe wall.

Csoma's Room Foundation, and then built over a period of years with village support (Figures 14.1, 14.2 and 14.3).

NGO involvement

The establishment of several active and well-connected non-governmental organizations in Ladakh, starting in the 1970s, has created an outlet for numerous mission-driven projects. These organizations provide valuable vehicles for outside funding and labor, as well as ideas and organizing. They occupy a space that is not entirely external, often hiring local employees and forging long-term relationships over a series of different initiatives and built projects. These groups bridge broader mission-driven goals and the specific needs of local stakeholders.

The organizations that have been active in Ladakh on climate-adaptive design projects include the Students' Educational and Cultural Movement of Ladakh (SECMOL), Leh Nutrition Project (LNP), and Ladakh Ecological Development and Environmental Group (LEDeG).[2] Many of these organizations have local leadership, or a combination of Ladakhi and international staff. They work as vehicles to support various specific campaigns, such as snow leopard protection, and can adroitly channel funding from an array of sources, such as the government, Indian corporations, and private external donors.

In scholar Cai Heath's study of the effectiveness of NGOs on climate-friendly development in Ladakh, village stakeholders noted that there is still a broad gulf between the types of projects available and the services that villagers need (Heath 2015). This report makes a case for even stronger communication between village stakeholder groups and the NGOs that initiate climate-adaptive projects. Many of these organizations have become fixtures in Ladakh over several decades, which enables them to continue to develop and support ongoing projects, long-term relationships with locals, and work within a well-known political, economic and environmental context.

Government support

The Indian government provides critical support to the region, seen through major programs in border security, road maintenance and construction, the ration system, schools, power generation, and other various campaigns. Ladakh's physical proximity to contested borders ensures that the government will invest heavily in various security efforts. The region has a much smaller per capita population than other parts of India, but garners a relatively large share of the country's resource allocation and investment. Programs active in Ladakh include the Watershed Development Programme, the Army's "Goodwill" Operation Sadbhavana program, The Ladakh Desert Development Programme (DDP) and the Ladakh Autonomous Hill Development Council (LAHDC).

Research in the region

The International Association of Ladakhi Studies (IALS) is a group of scholars and practitioners from a variety of disciplines working in Ladakh. The group convenes every other year at a conference, and maintains an organization for information dissemination. This organization has been a fundamental source of research about the region, providing a platform for professional networking and sharing of findings involving Ladakh. One of the primary challenges for the group has been the incorporation of local scholarship; without a regional university, scholars dealing with Ladakh necessarily come from other places. Over the years the group has worked to draw in Ladakhi collaborators, and to involve Ladakhi scholars who might have been trained at universities outside the region.

Benefits

Since the 1970s, NGOs, individuals, and design advocacy groups have visited Ladakh in an effort to share their knowledge. These experts have noted, among other things, the extreme scarcity of heating fuel, the relative thermal discomfort experienced in Ladakhi households in the winter, and the region's abundant solar access. Such encounters have given rise to new sustainable designs, such as the deployment of the trombe wall, improved insulation efforts, and the construction of attached greenhouses. A variety of new materials and technologies have been introduced, and entirely new public buildings realized. These collaborations have yielded new idea sharing in the realms of architecture, landscape and planning; leading to departures from vernacular practices both in terms of form and function.

Sustainability, too, has come to characterize the projects supported by allies in Ladakh. Many sponsored buildings interrogate the notion of sustainable development not in a general sense, but with regard to the specific environmental factors of the region. Whereas Ladakh's traditional buildings typically oriented to the south and reflected a massing to reduce heat loss, projects produced in collaboration with external allies tend to showcase many more visible sustainable design goals. For instance, newer buildings designed by visiting architects tend to incorporate insulation, trombe walls, large expanses of south-facing glass, and seismic framing, all of which were uncommon in early Ladakhi structures (Figure 14.4).

New design ideas, and the construction techniques that must be used to achieve them, have also become appropriated by Ladakhi builders, to become reimagined on other types of projects. Carpenters who participate in the development of a public-interest project sponsored by a foreign

Figure 14.4 The LEDeG campus in Leh showcases sustainable building strategies.

designer are much more likely to use these standards in future work, regardless of the sponsor or design agenda. Valuable new construction practices incorporating contemporary safety measures, new technologies, or environmental benefits, have gradually been added to the repertoire of normative building practice, as employees incorporate new training. This appropriation signals the stickiness of the design ideas imported from abroad, and also suggests the enduring impact of visiting design engagement in Ladakh.

NGOs, government groups, researchers, and visitors bring real value to Ladakh, in the form of investment, manpower, sponsorships, ideas, and technologies. A number of useful public buildings have been created with these partners, including schools, clinics, museums, and other types of institutional buildings. These projects, and the capacity with which external allies support them, provide a model for future climate-adaptive design efforts. Collaborations with allies often yield projects that combine social interests for the public good with other types of adaptive environmental goals.

In addition to bringing in new ideas and energy, foreign, NGO, and government sponsored projects also provide critical development capacity to the fledgling goals of Ladakhi communities. Free materials, labor, and design assistance can make an enormous difference in project conception and then completion (Figure 14.5). Access to crowdsourcing platforms and other creative funding mechanisms may enable Ladakhi communities to

Figure 14.5 A design review meeting with designers and clients in Zanskar.

leverage much larger donor bases. Without the financial and physical support of these organizations, the projects may not happen at all.

Limitations

As would be the case in any other rural contexts, Ladakhi communities have their own embedded cultural sensibilities, aesthetic loyalties and functional attachments that necessarily impact vernacular construction motivations and outcomes. Many of the building traditions in the region derive from a place-based wisdom developed over many centuries (Crook and Osmaston 1994; Norberg-Hodge 1991). In this context, differing ideas imported through pro-bono design engagement threatens to dismantle or challenge established modes of production.

In addition to the challenge of reconciling competing ideas about design values in Ladakh, the cultural context of the region provides another barrier to equitable power sharing in collaborative design engagement ventures. Many Ladakhi villages have been organized around a strong shared interest in maintaining collective harmony, and in general they may go to great lengths to ensure consensus (Mingle 2015). If Ladakhi cultural norms support acquiescence, rather than consultation and interrogation, collaborative partners must employ creative ways to engage honest community participation and feedback.

In Ladakh, development allies may need to work with minimal resources, and building norms that intentionally stem from the context. When new techniques and technologies are imported, often the training for local craftspeople and builders and owners doesn't get transferred. Moreover, the products developed in other places may not work in a different context. Wireless building monitoring systems, for instance, cannot operate without a reliable Wi-Fi internet connection, which is far from the norm in Ladakhi villages.

Finally, building projects in north India have their own set of logistical constraints. Projects must be carried out high in the mountains, where limited accessibility, extreme weather, and material scarcity define development. While labor is both affordable and available for six months every year, tools and technical skills are more difficult to manage. Transportation for both people and materials can be a major constraint, due to the poor condition of the roads when they are open, and the closure of major passes from October–May. Allies who are recruited to assist in development projects in Ladakh must work within these contextual and environmental conditions in order to be effective collaborators.

Moving forward

The vernacular buildings of Ladakh have been born out of centuries of trial and error, and thus they reflect time-tested solutions that respond to the challenging environmental constraints of the region. They are also socially, culturally and economically suited to the region. But perceptions about buildings and landscapes are changing, as new ideas are disseminated by visiting foreigners, researchers, NGOs, or through television, film, and the independent travels of Ladakhi people. In this sense, it is perhaps worth noting that construction techniques are not fixed in time; they evolve and change as the social, environmental and economic pressures shift.

While it is tempting for visitors to mourn the loss of old ways of building in a region that remains somewhat geographically isolated, the new efforts to adapt Ladakhi villages to a changing climate signals an intentional development course. Moreover, collaborations with external allies may also offer opportunities to improve quality of life factors. Indeed, visiting designers tend to import the high architectural standards learned in their home countries; integrating best practices for energy efficiency, thermal comfort, environmental stewardship, and building craft. They have literally brought in precision tools and high-tech building monitoring systems, new safety standards and factory-fabricated building materials. In so doing, these projects have helped to train a new cadre of local builders, capable of blending old and new styles.

However, in other contexts as well as in Ladakh, the influence garnered through financial and physical aid can shift the project considerably. Ladakhi people need to be able to drive the decision-making around projects that evolve in their space, and the sponsorship of others can get in the way (Chostak 2016). Almost twenty years ago scholar Martijn van Beek acknowledged this need, while also suggesting that Ladakhi development should occur in partnership with external (NGO, governmental) groups (van Beek 2000). Too often, van Beek notes, the region's development organizations

> are heavily dependent on outside funding, draw heavily on Western ideas and practices, and are remarkably hierarchical and bureaucratic, which has given rise to negative perceptions and suspicions among the urban and rural population, which in turn has proven to be a serious impediment to the organisations' work.
>
> (van Beek 2000, 264)

Development projects in the region have changed since then, and while this critique remains a legitimate concern for future projects, many of those organizations have since shifted their practices. In recruiting and then working with external allies, Ladakhi stakeholders have the opportunity to leverage a powerful source of support for realizing their intended visions.

Conclusion

Ladakh's long list of collaborative development projects showcase new forms of technology, enhanced sustainability goals, better fabrication standards, and contemporary material choices. Many of them build in opportunities for community engagement and stakeholder involvement, or even co-design processes. These landscape, planning and building projects assert new formal ideas while also incorporating principles drawn from traditional practice. In so doing, outside allies may effectively inspire the character and shape of Ladakh's villages; both in the public spaces that they commission and in future development where new design language might be emulated.

The sponsored projects brought about by allies from outside Ladakhi society may also bypass local norms and standards, and in so doing, call into question the ownership of a new vernacular form. After all, this neo-vernacular thinking has been brought in by outsiders; it is a gift that challenges both the character and ownership of Ladakhi space.

Regardless of motivation, in integrating best practices from abroad with contextually-appropriate design moves from the past, these new collaborative works may set the tone for one climate-adaptive design approach in Ladakh. Through new building techniques, materials, and forms, visiting designers have already had a significant impact on adaptive design work. As demonstrated by some of the projects in this collection, the local, traditional knowledge that has defined traditional placemaking ideas in Ladakh has begun to shift, ushering in new attitudes towards space.

In the rapidly-modernizing context of north India, where progress is increasingly measured by the physical manipulation of landscapes and buildings, external design practitioners can help locals envision a current, future-forward development agenda. The new structures conceived, funded, and built by allies also provide physical evidence that villages are in the process of shedding old traditions in favor of new, optimistic futures. In borrowing design ideals and aesthetic language from abroad, these constructs not only reinforce global notions of progress and improvement but also signal the village group's ascension into the modern world. In this sense, development trends in present-day Ladakh resemble the development trajectories witnessed in other parts of the world, where design language serves as both a symbol and tool for liberating a social group from outmoded trappings of the past. The creation of narratives of hope and future betterment are powerful tools indeed: as long as they are accompanied by equitable and responsible collaboration strategies.

Notes

1 For example, many of these new projects employ a functional space appropriation strategy of buffer zones, which create a thermal envelope for the building. While these spaces vary in terms of their design approach, aesthetics and program, they share the historic function of insulation through the use of the "thermal onion" concept.

2 A broader list of NGOs operating in Ladakh, with contact information, is available in the Appendix.

References

Aggarwal, Ravina. 1997. "From Utopia to Heterotopia: Towards an Anthropology of Ladakh." In *Recent Research on Ladakh 6: Proceedings of the 6th International Colloquium on Ladakh, Leh, 1993.*, edited by Henry Osmaston and Tsering Nawang, First Asian Edition, 21–8. Delhi: Motilal Banarsidass Publishers.

Beek, Martijn van. 2000. "Lessons from Ladakh? Local Responses to Globalization and Social Change." In *Globalization and Social Change*, edited by J.D. Schmidt and J. Hersh, 250–66. London: Routledge.

Brillembourg, Alfredo, Hubert Klumpner, and Patrick Coulombel. 2011. *Beyond Shelter: Architecture and Human Dignity*. Edited by Marie Aquilino. New York: Metropolis Books.

Charlesworth, Esther. 2014. *Humanitarian Architecture: 15 Stories of Architects Working after Disaster*. 1st edition. New York: Routledge.

Chostak, Stanzin. 2016. "Local Adaptation Strategies to Climate Change: Learning from Ladakh." *ILI Law Review*, Winter: 7–24.

Coulombel, Patrick. 2011. "Open Letter to Architects, Engineers, and Urbanists." In *Beyond Shelter: Architecture and Human Dignity*, edited by Marie Aquilino, 291. New York: Metropolis Books.

Crook, John, and Henry Osmaston. 1994. *Himalayan Buddhist Villages: Environment, Resources, Society and Religious Life in Zangskar, Ladakh*. Bristol: University of Bristol.

Devers, Quentin. 2017. "Charting Ancient Routes in Ladakh: An Archaeological Documentation." In *Interaction in the Himalayas and Central Asia: Processes of Transfer, Translation and Transformation in Art, Archaeology, Religion and Polity*, edited by Eva Allinger, Frantz Grenet, Christian Jahoda, Maria-Katharina Lang, and Anne Vergati, 321–38. Vienna: Austrian Academy of Sciences Press. www.academia.edu/39101653/Charting_Ancient_Routes_in_Ladakh_An_Archaeological_Documentation.

Frampton, Kenneth. 1983. "Prospects for a Critical Regionalism." *Perspecta* 20: 147–62. https://doi.org/10.2307/1567071.

Heath, Cai. 2015. "Climate-Friendly Development: Analysing Relationships between Community, Society and Government on Sustainable Technology Projects." *Ladakh Studies* 32 (January): 18–35.

Hyde, Lewis. 1979. "Some Food We Could Not Eat: Gift Exchange and the Imagination." *The Kenyon Review, New Series* 1 (4): 32–60.

Kahn, Louis, Paul Weiss, and Vincent Scully. 1953. "On the Responsibility of the Architect." *Perspecta* 2, 45–7.

Lo Bue, Erberto F., and John Bray. 2014. *Art and Architecture in Ladakh : Cross-Cultural Transmissions in the Himalayas and Karakoram*. Brill's Tibetan Studies Library: Volume 35. Boston: Brill.

Mingle, Jonathan. 2015. *Fire and Ice: Soot, Solidarity, and Survival on the Roof of the World*. 1st edition. New York: St. Martin's Press.

Norberg-Hodge, Helena. 1991. *Ancient Futures : Learning from Ladakh*. San Francisco: Sierra Club Books.

Nussbaum, Bruce. 2010. "Is Humanitarian Design the New Imperialism?" Fast Company. July 6, 2010. www.fastcompany.com/1661859/is-humanitarian-design-the-new-imperialism.

Prakash, Sanjay, ed. 1991. *Solar Architecture and Earth Construction in the Northwest Himalaya*. Sustainable Development Series 5. New Delhi: Har-Anand Publications in association with Vikas Pub. House.

Pyla, Panayiota. 2015. "Crisis Spins." *Journal of Architectural Education* 69 (1): 8–12. https://doi.org/10.1080/10464883.2015.987068.

Wakeford, Katie, and Bryan Bell, eds. 2008. *Expanding Architecture: Design as Activism*. New York: Metropolis Books.

Wickersham, Jay. 2014. "Code of Context: The Uneasy Excitement of Global Practice." *Architecture Boston*, Winter 2014. www.archdaily.com/590300/code-of-context-the-uneasy-excitement-of-global-practice/.

15 The role of design

Introduction

As people look for alternative ways to ensure stability and sustainability in the coming years, the disciplines of design and planning are well situated to provide useful ideas, technologies and direction. For example, designers are trained to envision new and creative responses to problems, relate those solutions through expressive drawings and digital media, and coordinate complicated projects involving diverse teams of stakeholders. Designers are particularly well suited to helping communities move through periods of transition, by offering transformative ideas.

According to educator Adrian Parr, design thinking also could be of use in the face of extreme challenges, such as for communities dealing with climate change or rebuilding after a major disaster. She suggests that "Because of their practical focus, the design disciplines-urbanism, architecture, planning, product design and graphic design-are especially well positioned to make a positive contribution to the problems associated with climate change" (Parr 2009, 148). Indeed, designers manage challenging production needs for the built environment, and balance many different factors in the pursuit of a single project.

Design practitioners can offer valuable resources and support to vulnerable communities. Author Victoria Harris notes that

> Peoples living in countries low on the United Nations Human Development Index are far more likely to live in unsafe, poorly built, poorly located accommodations, and are far less likely to have the resources to cope with or recover from catastrophic pressure.
>
> (Harris 2011, 18)

Designers can import fresh ideas and technologies, or expertise, to new areas. They can help communities cope with building challenges after a disaster, or in the wake of other stressful events. They can help communities project and plan, a luxury in many places – and prepare for a response to climate change.

Such a conceptual shift in the design disciplines also enables a more expansive and even altruistic form of practice. According to author Gautam Bhatia, "Unless architecture transcends its traditional scope, (i.e. working for the rich or for large institutional projects) architects will do incalculable damage to the environment and to the existing patterns of society" (Bhatia 2003, 23). This critical lens places responsibility on the designer to control for the impact of their work; it acknowledges that there is a central ethical component to design practice.

Climate-adaptive design

The functional demands of Ladakh's adaptive buildings and landscapes suggest a level of environmental engagement and potential for design innovation that could situate these interventions more firmly within the context of design thinking. One of the reasons that this work remains disciplinarily ambiguous is that many these design interventions perform, managing ecosystem services, for instance for climate change adaptation. However, just as the landscape urbanism movement at the turn of the twenty-first century opened up the discipline of landscape architecture to address issues of urban functioning, the adoption of climate-adaptive infrastructure into design discourse could highlight a shift in thinking about a more expansive and relevant role for designers.

Indeed, firms such as STOSS, in New York City, Bjarke Ingels Group in Copenhagen, and West 8 in Rotterdam each endeavor to link climate change challenges with the production of design work. In these offices, designers have been able to demonstrate the disciplinary value that they bring to the climate crisis, by producing evocative drawings, coordinating large groups of stakeholders, and delivering projects with benefits beyond efficient engineering. In her book *Hijacking Sustainability*, educator Adrian Parr suggests that "The combination of technical knowledge, practical focus, and creative experimentation of the design field means it is able to directly alleviate some of the debilitating effects natural disasters wreak on the lives of individuals, families, and entire communities" (Parr 2009, 8). Meanwhile, designers have much to learn from the knowledge and practice of scientists, engineers, and local farmers.

Current design activity in Ladakh

Most of the organic, informal, and bootstrapped climate-adaptive design strategies featured in this book were ultimately conceived of and initiated by local stakeholders. They represent a series of design hacks, referencing lived experience and undertaken with an inherent place-based wisdom. While much of the work highlighted in this book has not been produced in collaboration with designers, it nonetheless represents extraordinary design thinking. The informal design solutions that have emerged from the specific context and culture of Ladakh also offer real insight into both the opportunities and the struggles associated with mountain development.

Adaptive projects that incorporate design expertise, in the traditional sense, are limited. Artificial glaciers and ice stupas incorporate engineering design but few formal landscape architecture plans. Snow barrier bands, food security systems, tree planting and traditional solar development occur largely without the benefit of design methodology. Improved passive and active solar systems incorporate design engineering from professionals, largely within the building science disciplines. Strategies such as working with allies occasionally incorporate design professionals, but these interactions are limited.

Benefits

In Ladakh, external aid coming from trained designers may bring construction methods up to acceptable safety standards. For instance, the oversight of visiting architects can ensure that new schools, clinics, and other public buildings have the wall-to-ceiling connections and integrated rebar needed to withstand earthquakes. Designers trained to work in places where safety standards are high will likely maintain these standards in their aid work abroad.

Design ideas that have been specifically tailored to the region can draw upon science and technologies from other parts of the world, with specific application to the context of Ladakh. The book *Solar Architecture and Earth Construction on the Northwest Himalaya*, for instance, was published in 1991 by non-Ladakhi designers and then continued to inform development projects across the region in subsequent years. This book details solar building recommendations for Ladakh, produced by a group of building scientists and tested in several pilot projects across the region.

Limitations

When visiting designers come into a region like Ladakh, without the benefit of contextual and cultural awareness, they can also create design solutions that lack an appropriate fit. Aesthetic preferences differ across cultures, and so outside values and beliefs can be unwittingly pressed upon a new context. Similarly, materials and finances that work in other areas of the world might not make sense in Ladakh, and visiting designers may not fully understand these constraints.

One classic example of this tone-deafness in practice involves sanitation systems designed for Ladakh. Locals and most NGO aid groups recognize the value of retaining the Ladakhi dry toilet – a waterless composting toilet that serves as a vehicle for recycling much-needed human nutrients into the soil. But new designers to the area, especially the ones working on behalf of hotels in the area, automatically import sanitation technology from the rest of India. This causes the use of scarce water resources, limits function-ality during months when pipes freeze, and creates the problem of manag-ing waste removal.

Discussion

Performative projects, such as infrastructure or engineered landscapes, can be largely invisible to the societies they serve, despite the fact that many of these critical systems use enormous amounts of energy, tend to be central to the health of humans, and can fail spectacularly in natural disasters. As changing climates necessitate new types of human environments, the older, centralized models for infrastructure can give way to fragmented networks and small-scale interventions. Unlike single-use design projects, the inte-gration of multiple-use programming, performance, and ecological func-tioning for the large climate-adaptive projects of the future may even demand the expertise of designers.

The design disciplines specialize in the seamless integration of perfor-mance and experience by envisioning space in creative ways. In demanding resilient and sustainable design responses, climate change suggests a role for the design disciplines, where fundamental sustainability interests fore-ground new built work. Through climate-adaptive design interventions, environmental designers could effectively reposition their field to address the global climate crisis.

The case studies from Ladakh featured in this book reflect the low-tech, rural application of climate-adaptive design work that could become appropriated and formalized by the design disciplines. The deployment of small-scale design hacks – in essence strategic design interventions in the deserts of Ladakh – present an inspiring model for design practitioners to learn from. Within the context of the shifting design fields, this genre of climate-adaptive design work highlights the current disciplinary reposition-ing that embraces the topic of resilience and sustainability within design discourse. Design programs responding to climate change, for instance, share many of the key components of sustainable development, including an emphasis on performance, and the potential for aggregation, network development, and incremental growth over time.

During the course of the last several decades, designers have moved far beyond aesthetic goals to participate in the development of performative buildings and landscapes, earthworks, and infrastructural projects. In con-sidering climate-adaptive performance, ecological functioning and critical narratives in large-scale projects, the early pioneers of this disciplinary expansion effectively made the case for design practice in areas that other-wise might have been overlooked as relevant sites. Meanwhile, design the-orists supported this shift in practice by challenging designers to consider a new, synthesized approach, which included both infrastructural elements and change over time. In moving the discipline beyond the exclusive domain of human experience to assume broader goals for wildlife conservation, cultural preservation, ecological functioning, and the development of large-scale productive landscapes, the design disciplines have helped to usher in a new form of contemporary practice.

Moving forward

Many of the climate-adaptive projects in Ladakh are, at best, indifferent to the presence of human observers (Figure 15.1). However, the vision and expression that designers bring to understanding the experience of place could be put to productive use on such projects, where farmers, villagers and visitors spend time (Figure 15.2). Landscape architect James Corner considers this attention to the cultural, utilitarian, and ecological functioning of the site: "in terms of the retrieval of memory and the cultural enrichment of place and time; second, in terms of social program and utility, as new uses and activities are developed; and, third, in terms of ecological diversification and succession" (Corner 1999, 13). These three lenses for design thinking figure prominently in the emerging field of climate change design work, where designers link the essential services provided by infrastructure to the ecological functioning of surrounding environments and the cultural narratives that have come before.

Figure 15.1 The solar devices in this public landscape in Ladakh may be useful, but lack design thinking around place-making.

Figure 15.2 The *zing* in north Leh functions as a public ice skating rink in the winter months.

Examples of this design integration at the infrastructural scale include many of the ubiquitous spaces that humans use for recreation and agriculture in other parts of the world, such as levee parks, complete streets, and farming polders. In India, the *talaab* and the *maidan* stand out as important examples of multifunctional landscapes held in the commons, with varying degrees of design and planning (Mathur 1999; Nawre 2013). Could the Ladakhi women hockey players practice on a surface that also provides climate-adaptive irrigation benefits?[1] Might tourism expand to leverage the spaces produced by adaptive design? To overlay additional designed elements onto performative climate-adaptive projects would be to harness additional co-benefits, and in so doing, seat climate-adaptive work within the broader values of a society (Figure 15.3).

Just as multifunctional landscapes came to be recognized in the landscape and planning disciplines for knitting together ecosystem services and diverse human needs, climate-adaptive design projects could intentionally engage more heterogeneity than the single environmental objective. Scholar Alpa Nawre explained the idea of layering normative and situational acts in a parallel geography: water reservoirs found in India. In her case study on the *talaab* (ponds), ecological performance, infrastructural

Figure 15.3 A public meeting involving stakeholders provides a means of sharing design ideas and ensuring buy-in.

utility, and social interaction combine to produce a "laminate" space imbued with social and spiritual meaning (Nawre 2013). Similarly, some of the climate-adaptive design projects featured in this book apply a layered design approach involving a multiplicity of stakeholders and ambitions, and attendant complexity. These examples demonstrate some of the co-benefits that can come from multifaceted climate-adaptive design process and products, and in so doing, highlight new opportunities for the future of design practice.

In Ladakh, designers can capitalize on new technologies, ideas, and resources that would boost long-term resilience and sustainability indicators under the pressures of climate change. While adaptive projects present as important vernacular design responses to pressing environmental problems, their contribution to the design disciplines is perhaps less obvious. When installed and engineered by non-designers, do such projects still bring value to professional discourse? In many respects, adaptive projects have more in common with dams, levees, and agricultural terracing than the valued public areas so often associated with the design disciplines. However, these systems exist as a response to climate change, and while lacking formal design qualities, this emergent field could incorporate new experiential layers when framed by the professional history and aspirations of design thinking.

Although the relatively small number of outsider-sponsored projects undertaken in Ladakh can only begin to portray the character of design advocacy work abroad, they do reveal some of the new types of challenges and opportunities designers face when working outside of their home environment. The incorporation of design thinking in the context of Ladakh includes a combination of outsider and insider participants, each bringing their own design values. This pairing creates a unique challenge for Ladakhi development, in terms of ensuring equity and access, particularly in the creation of climate-adaptive design projects.

Designers working in Ladakh, particularly on climate-adaptive projects, need appropriate methods for useful, intentional, and conscientious community-engaged design assistance. Stakeholders need to be at the center of this work, holding positions of power and making decisions about how, and when, work is implemented (Chostak 2016). Recent research makes a case for tempering the neoliberal and homogenizing impulses that have heretofore characterized much foreign aid development, suggesting instead that practitioners focus on the individualized needs of places and people.

Conclusion

While the domain of designers has been traditionally differentiated from other types of engineering and infrastructural projects by providing spaces that are meant to be experienced by humans, this limiting perspective has expanded to accept many other forms of design engagement. This is a necessary first step for the incorporation of design thinking in climate-adaptive design and development. As climate change necessitates new types of space – across scales from the storage of food to the management of watersheds – designers could be of use.

Although design engagement has essentially been limited to external aid efforts in Ladakh, there is a case to be made for increasing the role of design in future climate-adaptive projects, and for including local knowledge and wisdom. Designers can bring valuable experience, interest and skillsets to projects that deal with the challenges of climate change. But these designers must be familiar with the context and culture of Ladakh, or even better, members of the communities in which they practice.

While this book addresses a number of projects that have been imagined, constructed, and managed by non-designers, this work might even be improved with additional professional support. Other projects explored in this book have been introduced by teams of visiting designers, with attendant challenges in terms of regional or local fit. The climate-adaptive design projects of the future could benefit from collaboration between local actors and design experts.

The adaptive projects in Ladakh highlight an opportunity for design engagement, and an obvious corollary: the responsibility of design practitioners to use their disciplinary training to address global climate change. The unexpected weather forces that will shape landscapes of the future will demand innovation and flexibility, and could benefit from the creative, systems-based approach of the design disciplines. Climate change could ultimately rewrite professional norms, creating a sea change in the types of projects that designers engage with in the future. Such a scenario would provide a new opportunity for designers to improve interdisciplinary collaboration, a means of influencing the work traditionally dominated by engineers and developers, and to have a say in the shape and structure of climate-adaptive space.

Note

1 *The New York Times* identified a need for ice hockey rinks in Ladakh in 2019 (Swift 2019).

References

Bhatia, Gautam. 2003. *Laurie Baker: Life, Work and Writings*. Delhi: Penguin Books. www.amazon.com/Laurie-Baker-Life-Work-Writings/dp/0140154604.

Chostak, Stanzin. 2016. "Local Adaptation Strategies to Climate Change: Learning from Ladakh." *ILI Law Review*, no. Winter: 7–24.

Corner, James. 1999. *James Corner: Recovering Landscape: Essays in Contemporary Landscape Architecture (Paperback); 1999 Edition*. New York: Princeton Architectural Press.

Harris, Victoria. 2011. "The Architecture of Risk." In *Beyond Shelter: Architecture and Human Dignity*, edited by Marie J. Aquilino, 12–22. New York: Metropolis Books.

Mathur, Anuradha. 1999. "Neither Wilderness Nor Home: The Indian Maidan." In *Recovering Landscapes*, edited by James Corner, 205–20. New York: Princeton Architectural Press.

Nawre, Alpa. 2013. "Talaab in India Multifunctional Landscapes as Laminates." *Landscape Journal* 32 (2): 137–50. https://doi.org/10.3368/lj.32.2.137.

Parr, Adrian. 2009. *Hijacking Sustainability*. Cambridge: MIT Press. https://mitpress.mit.edu/books/hijacking-sustainability.

Swift, Hilary. 2019. "'Hockey in Heaven.'" *The New York Times*, April 19, 2019, sec. Sports. www.nytimes.com/2019/04/19/sports/hockey/ice-hockey-india.html.

16 The future of climate-adaptive design

The trans-Himalayan mountain region of Ladakh is characterized by scarce annual rainfall, limited contact with the Indian subcontinent, and long, punishing winters. While the high-altitude landscape and extreme weather may appear to be too rugged and unforgiving for human settlement, Ladakh's tiny villages are comparatively lush, productive, and boundlessly hospitable. Time-tested techniques for the production of agriculture, food, and fuel, combined with new climate-adaptive practices, suggest a future in which Ladakhi people will not only carry on traditions, but in making contingencies, also thrive.

The climate is just one of many factors pushing change in Ladakh, and it triggers causal shifts in a host of areas. According to scholars Marcus Nüsser and Ravi Baghel,

> Regardless of the specific point of departure, research aims, theoretical foundations, and methodological designs of case studies, it is widely accepted that the livelihoods of most of the Himalayan population are under stress and adversely affected by continuous degradation or depletion of the natural resource base.
>
> (Nüsser and Baghel 2016)

Despite the relative stability and staying power of this society, environmental changes have also been met with intentional deflections, transitions, and pivots. This book introduces readers to a number of these adaptive design approaches, tested over the years, that enable cultural continuance under climate change.

In the high mountain villages of Ladakh, the self-sufficiency and relative autonomy of the group have historically been prized survival traits. Small, independent villages have operated as a collective unit, solving problems as they surfaced, working together to make decisions, and taking care of one another in times of need. Many of the climate-adaptive design ideas implemented in Ladakh acknowledge that villages have traditionally functioned as small, autonomous, self-governing populations, and in so doing reinforce age-old methods of group resilience. In both established traditions and in new design interventions, notions of radical autonomy underpin system functioning.

Ladakh

Indeed, self-reliance is fundamentally built into the culture and context of Ladakh. The region hosts some of the highest inhabited villages on earth, where for centuries sustainable design practices have been inextricably knit into agricultural systems and physical buildings (Prakash 1991). Sustainability in this landscape is a requisite; evidence of the region's extreme environmental pressures can be traced through many architectural and land use decisions. The region's climate and context has effectively created a vernacular pattern language for high-altitude villages.

It helps to look at Ladakh not just because it has experienced extreme isolation in the past and has thus developed elegant and extraordinary techniques for self-sufficiency, but because it also continues to face challenges today. In conceiving of Ladakhi modernity, one needs to focus on the present, local concerns of the region (van Beek 2003). Ladakh is located in a contested border zone with active military threats. It is a place defined by a climate in which it is possible to become physically cut off from the rest of the world for many months each year. It has a landscape that is difficult to negotiate, both because of altitude but also because of the steep terrain. And it is a place where tourism and economic development have begun to tax the underlying resource reserves in extraordinary and detrimental ways. These complications exacerbate the difficulties of this landscape, prompting design responses that reinforce systems of self-sufficiency.

On scarcity

It is instructive to look at the Ladakhi experience, where environmental resources have rarely been viewed as endless or even abundant, but could just be made to work through careful management. This attitude of scarcity is a fundamental ethic that has enabled Ladakhi society to thrive in spite of its difficult geography. As an essential principle, it stands in high contrast to the making of the modern state, where capitalism and its insatiable appetite for materials, energy, and land enabled unchecked growth. Under the pressures of climate change, population growth, and resource scarcity, this trajectory of development reveals underlying vulnerabilities. In order to even envision alternative models for development, it is helpful to study this ethic in practice in Ladakh.

After all, scarcity produces countless benefits. Behavioral economist Sendhil Shafir and psychologist Eldar Mullainathan posit a theory in which scarcity forces larger system efficiencies, by focusing "our attention on using what we have most effectively" (Mullainathan and Shafir 2013, 20). This idea connects directly to environmental management, where resource abundance often leads to waste (Mullainathan and Shafir 2013, 138). The relative scarcity of arable land, water, people, power, materials, and technologies in Ladakh may be responsible for limiting environmental degradation, solidifying social support networks, and spurring new areas of innovation.

However, not all of the effects of scarcity can be considered positive. While the condition of scarcity may actually increase attentiveness and effort, according to Shafir and Mullainathan, it also raises the costs of error, and provides more opportunity to make misguided choices. Psychologically, scarcity can reduce people's mental bandwidth, which leads to tunneling or an inability to consider the whole picture. Shafir and Mullainathan also note that scarcity "makes us less insightful, less forward-thinking, less controlled" (Mullainathan and Shafir 2013, 13). And in the case of design responses, scarcity can cause people to develop quick fixes rather than more thoughtful, premeditated, and well-funded interventions. According to Shafir and Mullainathan, "Scarcity not only raises the costs of error; it also provides more opportunity to err, to make misguided choices" (Mullainathan and Shafir 2013, 84).

Finally, scarcity can become a compounded effect over time. In Ladakh, limited supplies of critical resources have shaped the region's unique society and culture by requiring judicious environmental stewardship. The resulting balance between resource availability and use is a tenuous and precarious arrangement. Under climate change, additional environmental stressors disrupt these relationships. While some level of scarcity in the region has provided a real incentive for the careful management of the land and natural processes, excessive shortages could cause catastrophic failure to the whole environmental system.

Autonomy and change

Anthropologist and political scientist James C. Scott suggests that mountain people generally have become adapted to multiple modes of self-sufficiency due to their distance from state power. Moreover, he notes that in response to disasters, "Dispersal, routes of escape, and alternative subsistence routines must have been a part of the crisis repertoire" (Scott 2009, 163) of these mountain dwellers. Scott's work to understand the autonomous nature of mountain communities offers a lens with which to view the accomplished self-sufficiency that has structured traditional Ladakhi society. Even today the region has many effective systems in place for coping with adversity, which could be related to relative distance from state oversight and aid.

Unlike the top-down design solutions that may work well in large urbanizing contexts, Ladakhi villages have largely implemented climate-adaptive projects that are site-specific, small, and dispersed. While broad, overarching schemes for infrastructure, energy systems, and development projects may be necessary in large urban areas, in Ladakh these designs lack the very physical, cultural, social, religious, and economic features that might allow them to work. Instead, much of the development in Leh District could be considered contextually responsive, usually employing shoe-string budgets and available resources. However, precisely because of the region's relative autonomy, these projects also draw upon a foundation of building wisdom, a strong cultural identity of self-sufficiency, and in many cases, a willingness to engage in DIY projects that demand sweat equity.

While a fairly uniform vernacular village fabric remains dominant in most places, changing development modes threaten to erode many centuries of social, cultural and environmental stability in Ladakh (Daultrey and Gergan 2011; Norberg-Hodge 1991). The region has witnessed unprecedented change over the past five decades, through the rise of tourism, a burgeoning national defense program and an increase in new forms of government employment. Scholars note that this growth has also accompanied a concomitant deterioration in the practice of ancient farming techniques and agricultural reliance in Ladakh (Darokhan 1999; Mingle 2015), a reminder that this transformation brings with it both positive and negative undercurrents. While "planetary urbanization" (Brenner 2014) may open up new opportunities in terms of forms of employment, education, and entrepreneurship, it also suggests a shift away from the traditional cultural, social and physical fabric of Ladakh. This gradual globalization trend, combined with increased interaction with foreigners after 1974, has resulted in a growing stock of imported building types, tools, materials, and technologies.

The recent challenges to Ladakh's autonomy include a full spate of political, economic and cultural pressures. However, the really wicked problems in this area – the ones that impact the very viability of the region – stem from climate change. This book highlights the many creative coping strategies that subsistence agricultural farmers and enterprising engineers have devised to manage, cope, adapt, and even profit in this context. Among them, ice stupas, snow barrier bands, water reservoirs, solar architecture and greenhouses, tree planting and tree armor, food stockpiling, and artificial glaciers stand out as examples of bootstrapping climate change defiance.

The heightened public awareness around the devastating effects of climate change, brought into focus by an uptick in media attention, has stirred an outpouring of concern for the future of the planet. Planners and designers recognize that the future will likely be unstable and uncertain, making it difficult to prepare for using traditional planning strategies and techniques. However, as designers look for fresh ideas for engagement in this uncharted territory, the creative solutions developed by communities already experiencing climate change challenges stand out as valuable and relevant precedents. Journalist Jonathan Mingle suggests that communities on the front lines of climate change in northern India give the rest of the world a window into the future, where "The inhabitants' experience offers not only an early look at the kind of disruption likely to arrive in many other communities, but also some surprising lessons in human resilience" (Mingle 2009). While the adaptive designs highlighted in this book represent just a handful of the many ways that Ladakhi villages have approached changing environmental factors, together they start to build a series of strategies that could provide guidance elsewhere.

Transference

In the context of north India, shrinking glaciers and snowfields impact the amount of meltwater available to subsistence agricultural villages, and point to a future where climate-adaptive design solutions may be necessary for continued agricultural functioning. Novel design ideas that supplement and extend the traditional forms irrigation found in this formidable landscape are now within reach of local farmers (Nüsser and Baghel, 2016). The adaptive projects explored in this book present just a few of the creative ideas for improving the performance of Ladakh's dry desert landscapes.

Many of the climate-adaptive design strategies documented in Ladakh provide working examples of the type of interventions that could transfer to other cold mountain environments. These projects are characterized by their small size, dispersed systems, and responsiveness to local conditions. Design interventions perform within the autonomous structure of watersheds, villages, and households; they engage stakeholders in their development, from construction to maintenance; they augment existing systems to improve efficiency; and they are affordable, scalable solutions, accessible to rural farming populations. Many of the designs offer co-benefits: integrating multiple layers of use. Finally, these interventions are rooted in the communities they serve, connecting to the cultural, social and religious context, responsive to the climate and landscape, supportive of rural self-sufficiency and independence, and promoting affordability due to their scale and size.

Certainly, the applicability of these case studies to other parts of the world will vary considerably. It would be impractical to directly copy the creative climate-adaptive design ideas highlighted in this book to other places and situations. Ladakh is a highly complex environment, and many of these design ideas emerged out of a deep alignment with the place, climate, and culture. However, the work could be instructive to other regions, particularly those places struggling with the same environmental problems. How can practitioners and stakeholders borrow design ideas from Ladakh without being overly simplistic about the applicability of these projects? Is it even possible to anticipate infrastructural needs under changing climactic conditions? Perhaps the most applicable feature of the region is an attitude, characterized by an appreciation of traditional knowledge and wisdom, a willingness to embrace change, and boundless resourcefulness.

In many ways, it is helpful to look to Ladakh because it also is experiencing many of the types of constraints and challenges that we might expect to see in other places in the future. The broad themes of water scarcity, energy scarcity, and a lack of material resources could impact any other part of the globe. Similarly, challenges to mobility, or economic isolation, or the vulnerabilities of globalization could cause other parts of the world to seek a structure for society that more closely mirrors that of Ladakh.

The explicit use of climate-adaptive practices as a means of also strengthening Ladakhi livelihoods indicates an emerging tactic for environmental conservation. In this way, adaptive environmental solutions also highlight the value of uniting diverse and interdisciplinary stakeholders around a common social, cultural, religious, economic, or political cause. These efforts in Ladakh exploit layers of use and meaning, and stand out as models that could have practical application for other places as well.

As designers and planners prepare for change in other parts of the world, it is useful to look at the solutions and ideas that have worked in the high mountain context of Ladakh. Would artificial glaciers also work in Afghanistan, Baltistan, and Pakistan? Probably. Could snow barrier bands be applied in other parts of the world? Why not? We have already seen ice stupas appropriated in Switzerland and the GERES/LEDeG greenhouse model used successfully in Baltic States. In understanding the components of each of these design ideas, and the specific contextual response, it is possible to see how such thinking could be appropriated in other contexts.

If, indeed, instability will become the chief characteristic of climate change, adaptive resource management solutions will likewise need to become flexible, scalable, layered and dispersed. But as we have seen from Ladakh, in order to effectively deploy these varied approaches, pioneering solutions will need also need to fit in. Climate-adaptive design practices in Ladakhi villages demonstrate the value of integration with existing systems, social fabric, environmental context, religious interest and cultural identity.

The role of design

Design thinking can offer a solution-oriented method for addressing the major development challenges of our time. As communities develop plans for climate change adaptation, or restoring equity in the built environment, or envisioning new regenerative landscapes, the skills and processes provided by the design disciplines may become critical operating tools. If designers are to find meaningful ways to contribute to this work, they could benefit from the study of working precedents elsewhere. Models found outside of traditional design discourse abound, and they are reminders of the importance of environmentally, culturally, and socially responsive interventions.

Climate-adaptive design interventions in Ladakh provide an approach for design thinking with integrative, context-driven resolution. Could additional symbiotic relationships make their way into these design products? What would happen if a greenhouse were sized to also accommodate fruit trees, providing summer shade and additional agricultural opportunities? Could such a greenhouse also heat water for household consumption? Could that greenhouse be irrigated by an ice stupa?

These projects for climate change adaptation represent a new genre of design work that could have considerable implications for the design disciplines of architecture, landscape architecture, and urban planning. Ladakhi case studies highlight the work already accomplished by designers in climate change adaptation, as well as the potential that this creative thinking could have in other regions. Although many of these projects have been designed and implemented by farmers or engineers, they contribute to a larger body of climate-adaptive design solutions that suggest a way forward in the face of the unstable environmental pressures of the future. As designers look for opportunities to intervene in the climate crisis, the ongoing work found in northern India presents ideas for climate change adaptive design work to build upon.

Thinking big

Paradoxically, the independent spirit that has long characterized Ladakhi village life may hamper the region's ability to receive assistance from potential parties. In the face of climate change challenges, individuals and groups in the region could benefit from integrating additional forms of aid. The exchange of ideas, tools, products, expertise and technologies could augment the self-sufficiency that has long been practiced in Ladakh.

The assertion that humans can design their way out of environmental stewardship in the face of global climate change suggests some degree of arrogance. It would be hubris to believe that adaptive geoengineering experiments could miraculously solve the devastation brought about by global climate change (Goodell 2010; Fleming 2010; Michael Specter 2012). Yet, the climate-adaptive design interventions in Ladakh provide one example of the emboldened design schemes people are willing to try, given the dire circumstances that they face in their home environment. In this context, the disciplines of architecture, landscape architecture, and urban planning would do well to pay attention. Then, perhaps, the design process could become a useful tool in the face of climate change: eliciting creative responses and exploring survival strategies on a changing planet.

References

Beek, Martijn van. 2003. "Imaginaries of Ladakhi Modernity." In *Proceedings of the Tenth Seminar of the IATS*, 11: 163–88. Boston: Brill.

Brenner, Neil J., ed. 2014. *Implosions/Explosions: Towards a Study of Planetary Urbanization*. Berlin: Jovis.

Darokhan, Mohammad Deen. 1999. "The Development of Ecological Agriculture in Ladakh and Strategies for Sustainable Development." In *Ladakh: Culture, History, and Development between Himalaya and Karakoam*, edited by Martijn van Beek, Kristoffer Brix Bertelsen, and Poul Pedersen, 78–91. Denmark: Sterling.

Daultrey, Sally, and Reuben Gergan. 2011. "Living With Change: Adaptation and Innovation in Ladakh." *Climate Adaptation*. Available: www.yumpu.com/en/document/view/25089047/living-with-change-adaptation-and-innovation-in-our-planet.

Fleming, James Rodger. 2010. *Fixing the Sky: The Checkered History of Weather and Climate Control*. New York: Columbia University Press.

Goodell, Jeff. 2010. *How to Cool the Planet: Geoengineering and the Audacious Quest to Fix Earth's Climate*. 1st edition. Boston: Houghton Mifflin Harcourt.

Michael Specter. 2012. "The Climate Fixers." *The New Yorker* 88 (13): 96.

Mingle, Jonathan. 2009. "When the Glacier Left." *The Boston Globe*, November 29, 2009, sec. Home. www.boston.com/bostonglobe/ideas/articles/2009/11/29/when_the_glacier_left/.

Mingle, Jonathan. 2015. *Fire and Ice: Soot, Solidarity, and Survival on the Roof of the World*. 1st edition. New York: St. Martin's Press.

Mullainathan, Sendhil, and Eldar Shafir. 2013. *Scarcity: The New Science of Having Less and How It Defines Our Lives*. New York, NY: Picador.

Norberg-Hodge, Helena. 1991. *Ancient Futures: Learning from Ladakh*. San Francisco: Sierra Club Books.

Nüsser, Marcus, and Ravi Baghel. 2016. "Local Knowledge and Global Concerns: Artificial Glaciers as a Focus of Environmental Knowledge and Development Interventions." In *Ethnic and Cultural Dimensions of Knowledge*, edited by Peter Meusburger, Tim Freytag, and Laura Suarsana, 8: 191–209. Cham: Springer International Publishing. https://doi.org/10.1007/978-3-319-21900-4_9.

Prakash, Sanjay, ed. 1991. *Solar Architecture and Earth Construction in the Northwest Himalaya*. Sustainable Development Series 5. New Delhi: Har-Anand Publications in association with Vikas Pub. House.

Scott, James C. 2009. *The Art of Not Being Governed: An Anarchist History of Upland Southeast Asia*. New Haven: Yale University Press.

Author's note

It could be argued that Ladakh does not benefit from the appraisal of a visiting academic, and yet the stories of innovation in Ladakh deserve to be promoted. The innovation and ingenuity of the work underway and Ladakh has been underrepresented outside of the region, and still has much to teach people in other parts of the world. For this reason, this book seeks to explain Ladakh not so much in its own terms, but through the eyes of a visitor. And in doing so an effort has been made to make connections to other parts of the world so that the design thinking from Ladakh might be passed along.

My first introduction to north India was in 2012, when I was invited by a journalist to help out with a design/build project in a tiny village in Zanskar. The 1,000-year-old village, Kumik, had run out of glacial meltwater and was contemplating a move to a new riverside site. While touring the area and talking to locals about their approaches to climate change adaptation, I was impressed to see the variety of creative climate-adaptive work already underway. I began to research these techniques, drawing parallels to other parts of the world, and considering possible synergies. This book is the outgrowth of that research, and it is meant to showcase this work from a design perspective.

I hope that this book illuminates some of the interesting work happening in Ladakh, by individuals, engineers, NGOs, and government organizations. I hope that it will help visitors learn more about the projects they see in Ladakh. And I hope that it inspires readers to become more involved in climate change adaptation in their own backyards.

There are several ways to get involved in the region. The Appendix lists a number of different groups and organizations that are active in Ladakh, and have opportunities for visitors. Consider, for instance:

- Volunteering with one of the many NGOs in the area
- Participating in a workshop, or taking a class
- Joining a travel group with connections in the region
- Booking homestays with local families in Ladakh.

A large number of Ladakhi voices are accessible in these resources, and it is undoubtedly these local perspectives that will lead future research and development. Ladakhi-language publications, and Ladakhi scholars, have been woefully underrepresented in many scholarly venues, but this too is changing.

Finally, I would like to acknowledge the support that I have received from various editors in producing content for this book. Some graphics, images or portions of the following articles were re-published in this edition:

Clouse, C. 2016. "Design for Autonomy: Water Resources in Ladakh." In *Economic Modeling, Analysis, and Policy for Sustainability*, edited by Anandajit Goswami and Arabinda Mishra, 250–65. Hershey, PA: IGI Global.

Clouse, C. 2016. "Frozen Landscapes: Climate-adaptive Design Interventions in Ladakh and Zanskar." *Journal of Landscape Research* 41 (8): 821–37.

Clouse, C., N. Anderson and T. Shippling 2016. "Ladakh's Artificial Glaciers: Climate-adaptive Design for water Harvesting." *Climate and Development* 9 (5): 428–38.

Clouse, C. 2014. "Learning from Artificial Glaciers in the Himalaya: Design for Climate Change through Low-tech Infrastructural Devices." *Journal of Landscape Architecture* 9 (3): 6–19.

Clouse, C. 2018. "Seeking Appropriate Methods: The Role of Public-Interest Design Advocacy in the High Himalaya." In *Routledge Companion to Architecture and Social Engagement*, edited by Farhan Karim, 102–14. New York: Routledge.

Clouse, C. 2017. "The Himalayan Ice Stupa: Ladakh's Climate Adaptive Water Cache." *Journal of Architectural Education* 71 (2): 247–51.

An earlier version of the chapter on ice stupas appeared as an article called "Climate Adaptive Design: Building up Ladakh's Ice Stupas," in *Landscape Journal* vol. (38), no. (1), (Spring 2020). © 2020 by the Board of Regents of the University of Wisconsin System. All rights reserved.

Appendix

NGOs involved in climate-adaptive design projects in Ladakh

Groupe Energies Renouvelables, Environnement et Solidarités (GERES)
www.geres.eu/en/
Ice Stupa Project
http://icestupa.org
Ladakh Autonomous Hill Development Council (LAHDC)
https://leh.nic.in/lahdcleh/
Ladakh Ecological Development Group (LEDeG)
www.ledeg.org
(+91) 1982 253221
mail@ledeg.org
Ladakh Environment and Health Organization (LEHO)
(+91) 1982 252944
www.leho.in
Leh Nutrition Project (LNP)
(+91) 01982-252151
lnplehladakh@gmail.com
Ladakh Renewable Energy Development Agency (LREDA)
(+91) 019 82255733
http://ladakhenergy.org/about/
Students' Educational and Cultural Movement of Ladakh (SECMOL)
https://secmol.org/contact/
(+91) 1982 252421

Volunteer opportunities in Ladakh

Local Futures
www.localfutures.org/about/
SECMOL
https://secmol.org/get-involved/volunteering/
Snow Leopard Conservancy
http://snowleopardindia.org/volunteer-with-us.php

American educational groups operating in Ladakh

The BaSIC Initiative
https://basicinitiative.com
The Institute for Village Studies
www.villagestudies.org
Where There Be Dragons
www.wheretherebedragons.com

Indian aid organizations operating in Ladakh

Army's "Goodwill" Operation Sadbhavana program
Ladakh Desert Development Programme (DDP)
Watershed Development Programme

Academic groups connected to Ladakh

Association for Nepal and Himalayan Studies
http://anhs-himalaya.org
International Association of Ladakhi Studies (IALS)
https://ladakhstudies.org

Recommended reading

Himalaya, the Journal of the ANHS
http://anhs-himalaya.org/himalayajournal/
Ladakh Studies Journal
https://ladakhstudies.org/ladakh-studies-journal/
Stawa
www.facebook.com/stawaladakh/

Index

Page numbers in *italics* denote figures.